Christopher Wray's
GUIDE TO
DECORATIVE
LIGHTING

Christopher Wray's
GUIDE TO
DECORATIVE LIGHTING

BARTY PHILLIPS

With photographs by
NICK MEERS

Webb & Bower
MICHAEL JOSEPH

First published in Great Britain 1987 by
Webb & Bower (Publishers) Limited
9 Colleton Crescent, Exeter, Devon EX2 4BY

in association with
Michael Joseph Limited
27 Wright's Lane, London W8 5TZ
and Christopher Wray Lighting Emporium
600 King's Road, London SW6

Designed by Ron Pickless

Production by Nick Facer/Rob Kendrew

British Library Cataloguing in Publication Data

Phillips, Barty,
 The Christopher Wray book of decorative
 lighting.
 1. Lighting
 I. Title   II. Wray, Christopher

 621.32'28   TH7975.D8

 ISBN 0-86350-146-X

Typeset in Great Britain by Keyspools Limited, Golborne, Lancashire

Printed and bound in Hong Kong

CONTENTS

FOREWORD

Lighting has been my business for over twenty years and a wonderful Aladdin's lamp business it has been. The story has the quality of a fairy tale, since I 'fell' into lighting quite by accident.

After leaving school I followed various pursuits which included such diverse occupations as magician, dance-band singer, children's entertainer and actor. It was the last activity which led me into the world of lights. As an actor I was touring round England and Scotland, visiting different towns each week in repertory. I started to visit antique and junk shops and gradually developed a passion for collecting, beginning with coronation ware, old medals, clocks and oil lamps. One day my agent (who had just seen my stage performance!) suggested that 'you ought to get a side line'. The year was 1964 and the start of the 'swinging sixties': the Beatles, Twiggy and Carnaby Street. In London's Kings Road the first Antique Market was about to open. I took my chance and reserved a small space (10ft × 5ft). I put all my collectable goods on display, and set up shop.

The first item I sold was an oil lamp. I had seen others similar in a junk shop in Camden town so I rushed there to replace my stock and bought two more. These also sold immediately so I bought four more and again the magic seemed to work. Within a few weeks I had more old oil lamps on my stall then anything else. It gradually occurred to me that I ought to specialize in oil lamps. I therefore started to scour the country for unwanted lamps and glass shades, chimneys and spare parts and went from Lands End to John O'Groats in my search. The reader should appreciate that in the early sixties a few people in certain rural parts of Britain were still awaiting electrification and using oil and gas lamps to light their homes, especially in the depths of Devon and Cornwall. Even if the mains electricity had just arrived, most people still had in their cupboards 'this smelly old oil lamp'. They were delighted (and surprised) when I offered to buy from them what was in their eyes an ugly old functional light source of considerable inconvenience. I didn't agree, I thought they were beautiful and luckily so did my Chelsea and Kensington interior decorator customers.

Business flourished, and I expanded in the now famous Chelsea Antique Market. One day I came across an old Post Office at the bottom end of Kings Road. It was up for rent, and it seemed to me to be ideal to start my own shop specializing in lighting. I listened to advice from a friend of mine who had just finished a course at the London School of Economics. He had just completed a feasibility study of the possibility of success in opening a smart shop selling old oil lamps in a scruffy rundown part of Fulham. His research showed that I could not possibly succeed and that I would go bust within a year. I suppose, looking back, that I was obstinate, because, luckily for me, I did choose to open my shop and succeeded in taking the first year's rent within the first week. I decided to call the shop 'Christopher Wray's Lighting Emporium'. It was certainly an Aladdin's cave of lights, customers appeared as if by magic (I never had this success when I was performing tricks on the stage!).

Within a few months it became apparent that I could not keep up the stock level without organizing myself. The glass shades for the lamps were running out – I just couldn't find them anymore. I was forced to engage a small English glassworks to make replacement shades for me. They had hundreds of old moulds that hadn't been used for years, and they had the skill and personnel to produce for me exactly the models I required. They couldn't believe that anyone still wanted these old fashioned shades, and when I asked them to make them in different colours (instead of just white) they thought I was mad. But how lucky I was, I had almost by accident untrapped a demand which had started as a trickle and was now turning into a torrent. At the same time people who still used oil lamps began to hear of this shop which specialized in the selling and repair of old lamps. My stock of lamp chimneys for instance had grown from just a few models to over 120 different shapes and sizes. You cannot just put any chimney on any lamp. Each burner is different and requires a specific chimney. I was obtaining these old chimneys (some went back to the last century) from small country ironmongers who had them buried in their back sheds. 'Thers no demand' they told me and were glad to sell them to me and clear the space.

After a year I bought the shop next door in the Kings Road and enlarged the range of lamps. The original shop was turned into 'the Lamp Workshop' where I sold all the replacement parts for oil and gas lamps. I found an old building yard behind the shop and set this up as a workshop to repair and restore the old lamps. Until this time this work had been carried out at the back of the shop in view of the public. I was gradually engaging more and more colleagues around me to help run the growing business. I was still accepting acting jobs from my agent – mainly TV small parts. When I was offered a larger part in a programme called *Emmerdale Farm*, which was filmed in Leeds, I decided to call it a day as an actor and remove my picture from the Actors Directory. I just couldn't do both things and by now my lighting business was doing really well.

In the late sixties I was brought my first Tiffany lamp from America. I had never seen anything like it before and fell in love with the vibrant colours and graceful design. I decided to manufacture my own range of lamps based on the Tiffany style. I needed to obtain a source for the opalescent glass used in their manufacture. After much searching I found an original source which I still use today. The glass is hand mixed and especially made for use on lighting fixtures. We hand band each panel and the whole lamp is assembled by hand using the original techniques developed by Louis Comfort Tiffany at the turn of the century. After producing quite a range of these lamps I acquired other premises across the road, and engaged a few more personnel.

At about this time a colleague suggested that we open another shop and our first one was in Ireland at the Old Kilkenny Theatre. This was followed by Bristol, Birmingham and others around the country. All these shops needed a lot of stock and simultaneous with opening the shops I enlarged the workshops to a factory, and in fact took over an old manufacturing business in Birmingham (the home of brass). Here I found the skill and workforce to reproduce the lamps that I now sell. Most of them are still made using the original tools and dies and to the original designs which I have collected over the years. I don't really make reproductions but simply recreate the originals.

Today in London's Kings Road, I am embarking on the most ambitious plan so far. I have bought from the Gas Board a large piece of land opposite my original Lighting Emporium. A new factory, warehouse and one enormous shop to house my complete collection of decorative lighting will be built there. My customers will be able to purchase everything in lighting under one roof, instead of going from one shop to another.

And now you, the reader, sit back and enjoy the pictures and text of this super book.

Christopher Wray

INTRODUCTION

Lighting is one of the most important contributions to human well being. We are enchanted every day with the thousand ways direct sunlight and natural daylight show up colours and emphasize the contrasts between light and shadow and the shapes and textures in nature. It is this variety and changeability which interests and delights. And it is this variety together with the practical provision of necessary levels of illumination, for seeing to do jobs and living safely, that we should try and bring into our homes.

The eye is a marvellous instrument. It can adapt itself to an enormous range of light levels. It will automatically make the continuous changes necessary for good vision under changing circumstances. In fact, eyes and brain really need changing circumstances in order not to become bored and find their vision monotonous. However, that doesn't mean that we need to make things more difficult for the eyes than necessary. Practically speaking, the contrast between the amount of light shining fully on the pages of a book or other piece of work and the light shining in the immediate background should not be too great, or the eye will have a difficult time adapting to the contrast. You will see, if you try to work in a darkened room at a brightly lit desk, how hard this is. On the other hand, you do want some difference in light level between what you are concentrating on and what is going on round about. In practice, you will find that about three times the light level on your work than in the background is the most comfortable. From the age of about forty the human eye needs more light to see clearly, and this should be borne in mind when planning lighting where there are older people in the house.

Colour is an important aspect of lighting. Yellow or white light on the whole gives a friendlier light than the blue or grey tones, which is one reason why the GLS bulb is still in favour since it shines with a distinctly yellow light. Lamp shades can be used to add colour to the lighting and you can do quite a lot by contrasting or matching colours of both lamps and shades with the fabrics, wallcoverings and floor coverings of your home.

The key to most good home lighting is flexibility. Many low level lamps of different kinds will offer more effective, practical and satisfactory lighting than one or two high level sources. This gives much greater opportunity to alter and vary the lighting and to co-ordinate it with the rest of the decor.

It is now, more or less, generally understood that a home needs good task lighting for specific jobs around the house: for reading, cooking, sewing and so on. It is generally accepted, too, that accent lighting, highlighting, background lighting and fun lighting are also desirable. But these are difficult to get right. It is with these that this book mainly deals, concentrating on lights which are decorative in their own right, as well as those which create a decorative impression in their surroundings.

Finally, no matter what anybody else says, personal taste is more important than any theory about lighting. In the long run, you are going to live with what you choose. This book does not attempt to tell you what is right or wrong, but only to offer ideas. Some of them may appeal to you and some may not.

NB In the lighting trade a 'bulb' is something you plant in the garden! The glass globe which contains the light source is known in the trade as a 'lamp' and the fitting is a 'luminaire'. However, these are not terms normally used by most people at home or when shopping and so, throughout this book, 'bulb' is taken to mean the light source and 'lamp' means the fitting.

Opposite
This is the bare electric light bulb, pretty much the same as when it was invented by Edison (in the USA) and Swan (in England) in the 1870s. More than a hundred years later it is still the light source on which most home decorative lighting is based.

HISTORY OF LIGHTING

Sight is one of our most precious senses. History shows that
from earliest times much energy has gone into finding cheap,
efficient ways of providing light and into embellishing and
decorating the light source. Only in the last fifty years or
so has light become an integral part of the interior scheme of
a home.

It is surprising in our switch-on twentieth century that our grandparents and in some cases even our parents used to be sent off to bed clutching a candle whose flame guttered and threatened to go out, engulfing the child in blackness. We have become so used to summoning light at the flick of a switch, it's hard to imagine the darkness and the fright. Yet from prehistoric times right up to the nineteenth century, producing light was difficult, dirty, cumbersome and far from cheap. Edison, that genie of the lamp, didn't patent the electric light bulb until 1879.

The first artificial light employed was firelight, which was used primarily for warmth, cooking and protection from wild animals. The light was an added bonus. Brands plucked from campsite fires made torches which could be carried around; resinous wood was best because it flamed up to give a really good light. In fact right up until the twentieth century, fire was the main source of lighting in the home for more than half the world's population.

The development of lighting has always been through solving the problems of cost, efficiency and convenience. Anything flammable will make some sort of light, so practically everything cheap or close to hand has been tried in various forms. Oils have been expressed from animals and fish, which must have smelled disgusting, the seeds of grapes, tea, cabbages, olives and a good many other things besides. Throughout the centuries interior light has been provided by flame torches, tallow vessels, oil lamps, wax candles, paraffin lamps, gas and electricity more or less in that order. All were luxuries in their time, though few were really efficient or cost effective.

There was a constant search for an ideal fuel, which wouldn't smoke or smell and would give a good bright light. Tallow was generally made of animal fat or grease. When burned for light, it smelled to high heaven but then, people were accustomed to strong smells in those days. It was also useful in the form of candles and tallow-dip. The early tallow lamps gave a feeble, flickering light, conjuring up other-worldly shadows and terrifying shapes. In the western world, stone lamps were the basic form of lighting until the Middle Ages, and cresset (hanging) stones were used to light church porches and monastery cloisters. Shells which had natural spouts were also used at first. Later, people began to make the vessels for their lamps and stopped relying on finding suitable ones lying around, but the earliest fabricated lamps were made just like shells.

The classical 'day' was governed by the sun and people got up at dawn to use every moment of light and went to bed at dusk. Lamps were used sparingly except for study or on special occasions. The evening meal was begun before the sun set and only the rich went to 'all-hours' banquets, since in the absence of street lighting, they needed slaves with torches to guide them home. As one can easily imagine, 'lamp black' was a problem. Slaves had to wash down the decorations and statues after every banquet when light had been lavishly used. Ancient Roman interior decorators actually advised against decorating a room with frescoes and stucco reliefs in winter apartments because they were so difficult to keep clean. They recommended instead polished black walls with ochre or vermilion panels, which would hide the dirt.

An extraordinary amount of inventiveness has gone into lamp design through the ages. The royal tombs at Ur were found full of gold, silver, copper and alabaster lamps. Perhaps these were the first purely decorative lights ever made – ironically placed where no living eyes were supposed to see them. Sri Lankans used to suspend bird shaped, hollow, gravity-fed lamps from the ceiling and keep the lamp bowl full at its feet. The lotus lamp with separate brass or copper petals, was like a closed lotus bud when not in use, opening up like a flower to reveal the reservoir.

There are hundreds of carved relief designs of gods and goddesses, gladiators and mythological subjects, such as the jolly god Silenius with an owl sitting on his head between two huge horns, each of which supports a lamp stand; or a flower stalk growing out of a circular plinth with snail shells hanging from it by small chains, each shell holding oil and a wick. All this for what could only have been quite feeble light by today's standards. However, such sumptuousness is rare compared with the vast quantity of pottery and lead lamps by which the ordinary Roman citizen lit the twilight hours.

Some very ingenious lamps have been concocted using animal fat. Newfoundland fishermen used dogfish tails stuck into a cleft stick and American Indians did the same with 'candle fish'. In the Shetland Islands stormy petrels with wicks thrust down their throats were burned for light, and other oily carcasses (penguins for instance) have been used with wicks drawn through them. Much later, British soldiers in the First World War improvised lamps from sardine tins and, hungry as they were, used the oil from the sardines as fuel.

'Lamp black' has always been a problem when dealing with fuels, except electricity. Smoke catchers were used for paraffin lamps and are still made for people who use paraffin. They prevent the whole ceiling from becoming dirty but need to be unhooked and washed out occasionally. They look attractive too.

These traditional oil table lamps are all available today: a hand oil lamp which will see you up to bed without guttering out, a brass Corinthian column with a cut crystal glass fount and a lamp with a black ceramic base, a china fount and a tulip shade.

Left
A modern paraffin lamp, designed to use white paraffin, produces practically no smoke and gives a gentle light. It is used here as a cheerful welcome in the window of a small country cottage, though it is elegant enough to take the place of candles on a dinner table.

There were many versions of the spout lamp, in which the top was enclosed and the wick inserted into a long spout rather like a teapot. Some were double spouted like the Lucerna lamp from Italy which had a central stem passing through the middle of the reservoir so the light could be adjusted. They were not always very stable and it took the sensible Dutch to weight their lamps with sand to stop them tipping over. A reading version of the spout lamp could be raised and lowered on a central brass column and might have three flame spouts, and a brass screen to reflect light onto the page. The Scottish 'Crusie' spout lamp was used for centuries by the Celts and introduced into America by European settlers in 1620 as the Betty or the Phoebe lamp; it could be placed on the floor or suspended from the ceiling on a hook and used fish oil for fuel. Some had a double bowl to catch drips – you can be sure the canny Scots would not waste anything – and some could be tilted to catch the last dregs of precious oil.

By the early seventeenth century naphtha was being distilled from coal. This gave a brilliant light, and was useful for the outdoor lighting of large areas. The Holliday lamp, which burned naphtha, was developed and used for street markets for about one hundred years. Most European lamps upto the eighteenth century had side protruding spouts, but lamps with a central wick are more convenient. Benjamin Franklin in America thought to fit a central wick with a double burner, giving double the amount of light. (Three burners did not, unfortunately, triple the amount of light.) In the nineteenth century the luckless sperm whale was found to have a cavity in the head which yielded a huge supply of spermaceti, used for cosmetics and also for lighting. 'It flameth white and candent like camphire – some lumps afford a fresh and flosculous smell' said a writer of the time. When whale oil prices rose, lard became the popular fuel. In 1830 whale and lard were both overtaken by the introduction of 'camphene', a highly volatile mixture of turpentine and alcohol which turned out to be a dangerous explosive, causing a number of disastrous fires and much loss of life, although caps were placed over the lamps when not in use to prevent evaporation. Paraffin eventually ousted all other similar fuels.

Opposite
Another look to the past, with this small, traditional column lamp with a white fount. A white, slightly frilly shade shines down on and complements the elegant white ceramic swan, the patterned screen and the somewhat sumptuous background. Such lamps are available for use with paraffin or electricity.

Candle power

Primitive oil lamps began to go out with the fall of Rome when it became difficult to get the olive oil from the south of Italy. The northern vineyards, used by the rapidly spreading Christian church for growing grapes for liturgical wine, also had many bee hives and the wax was ideal for candles. These had advantages over oil in that they could be self supporting and lent themselves to much more intricate holders, candlesticks and candelabra.

In medieval times candles and their fixtures were usually kept in a niche by the fireplace or on a chimney ledge, where they could be lit from the embers of the fire – an example of early ergonomics. Upper class women (who could afford it) often kept a night light burning in their bedrooms all night. In fact, it was common to be 'lit' to bed by a servant, and husbands and wives used to keep the servant standing with the light while they read in bed. Eventually, the poor thing would be allowed to totter off with the stub of the candle to his/her own bed. Kings and princes had their way lit by candle bearers, in lieu of today's red carpet. In fact, it was an honour to 'hold up the candle' for someone. Candles were particularly fashionable at banquets where servants were used as human torch bearers. At the notorious Bal des Ardens, one of the torch bearers set fire to the costumes of the dancers causing tremendous uproar and panic. Several people died and Charles IV of France is said to have been driven mad with the horror of it all. Since servants were obviously dangerously unreliable as torch bearers, metal candle-

Candles became the mainstay of early lighting. They had the advantage of 'solid fuel' which was self supporting and could sit on a candlestick. These are examples of early hand lamps which include a lacemaker's lamp with a large reflector to direct light onto the work, a candlestick with a glass shade, a table paraffin lamp, a miner's lamp and a tall candlestick.

sticks began to be fashioned in the shape of standing men or horsemen, and disembodied arm sconces were fixed to walls.

Medieval pictures of interiors often show candles in holders or held in the hand – or sometimes just freestanding rushlights or tapers. Rushlights were made out of bleached and dried pith of rushes and 'the scummings of the bacon pot'. There was also the 'rat de cave', a roll of waxed wick looking like a ball of knitting. Special holders called 'wax jacks' were later developed to hold them.

In 1710 the wily British Government imposed a candle tax and it became illegal to make them at home. Officials in the king's household were often paid, in part, with candles – candle ends were the highly unofficial and frowned on perk of the royal servants.

Candlesticks

It is possible that the poor used wooden candlesticks, but of course wood could easily catch fire and was difficult to clean, so brass and iron were more popular and have certainly lasted longer. Simple candlesticks and candelabra diversified in the fifteenth century into sconces on swivels at the chimney piece and chandeliers. These were sometimes on a counterweighted pulley so that they could be pulled down and refilled on the rise and fall principle used nowadays for electric dining room and kitchen lighting.

Candlesticks themselves became both more ornate and labour saving, often made with spiral or ratchet winders so that, as the candle burned down, you could raise its base and keep the light at a constant height for reading and writing. In the seventeenth century the upper classes in Britain used silver for almost any kind of implement and furnishing in their homes. Silver candlesticks varied from the baluster shape to clustered tubular shafts, then to the true classic columnar form. This was when candelabra (silver candlesticks with two or more branches) began to appear. A reflecting plate of brass or copper was often nailed to the wall behind a candle to increase its refulgence, and light brackets were attached to mirrors to produce a similar effect. Rock crystal chandeliers were much admired, partly because they were so expensive. Only slightly less expensive were crystal glass chandeliers from Milan, Bohemia and France. Unparalleled heights of lavishness and design were reached with candlesticks carrying pendant drops of glass. So much so that by the nineteenth century candlesticks were designed as ornaments in their own right and often didn't even have sockets to hold any candles.

By the middle of the nineteenth century artificial light had transformed the lives of all but the very poor. Dinner now moved into early evening rather than late afternoon (perhaps specifically to show off the candelabra) and a cold luncheon was introduced in the middle of the day to keep everyone going. Although candles were convenient, they caused many fires. Hogarth's print of a short sighted man reading too near a candle is a typical scene of the time, and one man actually set fire to his hat while reading by candlelight.

Lighting 'on tap'

Gas and electricity

Gas fuel was used by the Chinese very early on, being brought up from 2000-foot wells and transported through bamboo pipes. But the first gas used for domestic lighting in the west was made from coal and not discovered until the nineteenth century. In the late eighteenth century people were still struggling to read by the light of wax and tallow, when Philippe Lebon managed to light and heat his own Paris home with gas piped from room to room by tubes buried in the plaster. (His experiments unfortunately came to an end when he was attacked and killed by muggers in the Champs-Elysées in 1804.) More or less at the same time, William Murdock succeeded in lighting his home in Cornwall by distilling coal in an iron retort. By 1802 he had put up gas flares at each end of the main building of his employers, engineers Boulton and Watt in Birmingham. Murdock's system was marketed by Watt's firm and 900 gas lights were installed in a large cotton mill in Lancashire. In January 1807 Pall Mall was lit by gas lights and in 1917 street lighting was installed in Baltimore.

Although gas lighting transformed nineteenth century life, early gas lamps provided only a modest yellow light and were so smelly they made the atmosphere in the room hot and oppressive. The early lamps were simply holes in gas pipes, but by 1893 gauze gas mantles were placed round the flame, burning white hot, making a much more satisfactory incandescent glow and increasing the light output. Later still, of course, the beautiful Victorian glass shades were provided which gave gas lamps their peculiarly pleasant and gentle quality.

Opposite
These are elegant gas lamps very similar to those installed in Pall Mall in January 1807. They still stand outside St James's Palace, and are equipped with the original gas mantels although the actual lighting has been converted to electricity.

Right
An old fashioned glass lamp shows how electric light can bring coloured glass to life.

But even as gas was being developed, people were experimenting with electric light. Humphrey Davy first produced electric light in 1808 in the form of the arc light. In 1877 London's Victoria Embankment became the first street to be permanently lit by electricity. Arc lights were, however, always far too bright and harsh for use in the home. It was Edison who, seeing a display of arc lights, realised that what was needed was for the electricity to be sub-divided so that the light could be 'diluted'. He turned his attention to the incandescent filament bulb, 'the less promising line of en-quiry'. Unlike anybody else, Edison envisaged from the start, not just the bulb, but a whole lighting system which could compete with gas lighting. Edison in America and Swan in Britain were hunting the incandescent filament bulb at the same time. Swan showed his model to a meeting of the Newcastle Chemical Society in 1878 but couldn't actually demonstrate it working because he had burned out the fila-ment in a lab test. He did give a well received demonstration in January 1879, but by that time Edison was working on his version. At that time Swan formed a public company, but in October Edison kept a lamp burning for thirteen hours and patented his bulb, something Swan had failed to do. Edison's first bulbs sold for $2.50 each and he opened his first electric light factory in 1880, closely followed by Swan who opened his outside Newcastle in 1881. Edison unsuccessfully sued Swan over infringement of his British patent. Eventually they came to a business agreement and formed the Edison Swan Electric Light Company for manufacture and marketing of light bulbs in Britain. Today Edison's screw bulbs and Swan's bayonet fit-tings are still manufactured and sold.

Fluorescent light was introduced at The World's Fair in New York in 1939. The inside of the tube of a mercury lamp was coated with phosphors and ultra violet radiation making these fluoresce and providing light. The colour depended on the type of phosphor chosen. Fluorescent tubes give about four times as much light per electricity consumed as incandescent bulbs and they last longer. They have become highly popular as working lights, and for chain stores because of their shadowless quality and the good colour reproduction they achieve.

The most exciting recent development in lighting is without doubt the development of the low voltage bulb. Low voltage bulbs operate on a supply of 6, 12 or 24 volts (rather than a mains supply of 240 volts in UK and 100 volts in USA). Most low voltage bulbs are very much smaller than conventional bulbs and almost any fitting can be made in a low voltage version, so their potential is enormous. Many operate on the tungsten halogen (or quartz halogen) prin-ciple, are very economical to run, sometimes producing two or three times more light for the equivalent wattage in a standard bulb. They generate less heat, and have a life of 2000 or more hours. Some are simply bulbs, like minia-ture GLS bulbs; others have an integral reflector round them. They are becoming rapidly more widely available, in many beam angles and wattages. All low voltage must have a trans-former to lower the voltage from the 240 or 100 volts of the mains. Many lamps are designed with individual transformers incorporated in the base of the lamp, some are incorporated in the bulb. Better still is to have a 'remote' transformer housed in a cupboard, from which you can run several low voltage lamps, but you will need specialist advice on this.

Opposite
These two little table lamps are good examples of how lamp shades retained the look of those used for gas for some time after electricity took over as the main source of lighting in homes. It was some time before anything really different and innovative was designed for electric lights.

HISTORY OF STYLE

Once electricity became available designers pounced on it as an exciting new way to add to a home's furnishing style. At first fittings imitated those of Victorian gas lamps, but from the 1920s on designers became aware of the tremendous decorative potential and lighting was incorporated into the fashions of the time.

Art Nouveau

In design terms, at the end of the nineteenth century, things suddenly began to happen very fast. Technological and scientific advances seemed to offer limitless possibilities and manufacturing industries offered jobs with good money to many more people. A whole new, affluent class emerged who had money to spend; objects were mass produced and department stores sprang up to sell them. As more people were able to afford nice homes and furniture, they wanted to buy interesting light fittings to go with them.

This coincided beautifully with the birth and development of electric light and the incandescent bulb which liberated designers from the constraints of other fuels: the dirt, inconvenience and storage problems. Electricity and the 'piping' of lighting all round the home meant that lamps could become an integral part of a home decorating scheme, and Art Nouveau was the perfect style to celebrate the new freedom and promise. There were three recognisable themes to Art Nouveau: the flower, the woman and the combination woman-flower. This was the time socially when women were being freed from their terrible whalebone harnesses and stifling bourgeois respectability. Pianos were built in Art Nouveau taste with piano candles. And although Queen Victoria may have favoured enormous glass droplet chandeliers in the new ballroom at Buckingham Palace, people in their own homes were installing things far less formal and traditional.

The new movement was epitomised by Art

Stained glass was prevalent in homes in the twenties; in the glass around front doors and in the front windows of many terraced houses. It used natural light to create sometimes spectacular effects like this large window advertisement which epitomizes Art Nouveau's influence on lighting with its voluptuous 'lady with the lamp' and the sinuous curves of the overall design.

Nouveau woman – flowing hair, large hips, generous proportions and clinging clothes. Lamps were made in the form of young girls personifying electricity. Their loose hair floated about the head; they wore transparent robes and held torches in their upheld hands to herald the new century. The undulating motifs were found everywhere; on book covers, title pages and chapter headings; on furniture carvings, glass, posters, stained glass windows, scarves, silver; on teapots and jugs, vases and hair brushes, and of course, lamps. Art Nouveau was eminently suited to lamps, where the electricity supply was brought up through the flower, like sap, to the bulb which gently lit the flower or bud, producing some exotic effects. The light source no longer had to face upwards

to carry the fuel, but could be directed up, down, sideways; any way at all. Orchid, poppy, cyclamen, honesty, or whole electric bouquets with a whorl of leaves or petals as reflectors were created. Art Nouveau was intended to unify and harmonise interior decoration, and lighting fitted perfectly into this idea. There was to be no more Neo Classical, Renaissance, Baroque, Rococo, Empire, Louis XV. Everything had to harmonise and 'match'. More light fixtures were designed at this period than would have been normal for any other. In fact every architect and designer seems to have designed at least one table lamp, chandelier, candlestick or wall bracket. After all, every room now had to be lit. Art Nouveau was aesthetically superb but highly impractical. Philippe Wolfers in Belgium

The electricity supply is brought up through this figure, which is a cross between a merman and an angel and which holds the brass stem in its hand from which emerges an opal globe. The mixture of heavy decoration and flowing line shows the transition from Victorian to Art Nouveau.

Frilly glass lampshades are a carry over from gas lamps. They blend well with the stem-and-leaf design of the double wall bracket, which leads into the flower-like lights. A sixties version of Art Nouveau wallpaper adds to the general early effect.

designed lamps using the plique-a-jour effect to make the enamel 'eyes' of a bronze peacock glow like jewels; Tiffany trapped light in the opalescent glass to highlight its luminosity, rather than allow people to read by the light.

Numerous materials were used for light fittings, especially bronze. Small foundries discovered that new homemakers wanted more and more objects for their homes and started to run serial bronzes to meet the demand. They would negotiate copyright with well-known sculptors, who could then earn royalties from mass production. Their sculpture was adapted into lavishly decorated mantel sets, ashtrays, epergnes, inkwells and lamps, using bronze, pewter, ormolu and spelter. Motifs for light fittings were the same range of themes as for other interior work. Some went for the botanical, some for the more austere and architectural, playing with vertical and horizontal lines; Parisian designers went for the whiplash curve and the curvaceous.

Tiffany and glass

Nowadays, the name Tiffany has become a generic term for all lamps and fixtures using coloured leaded glass and many people are not aware of what the word 'Tiffany' actually means.

Louis Comfort Tiffany, who gave his name to Tiffany lamps, was trained as a painter and then became one of the world's first professional interior designers. He had three great enthusiasms: the use of nature in art, oriental art and stained glass, and his inspiration was very

often William Blake. He started by experimenting with glass as a medium for his painting and followed that up by using it in his interior schemes for freizes, one of which was a panel of egg plants and gourds in opalescent glass for the dining room of rich New Yorker George Kemp. He redecorated the White House in 1882 and gave a touch of oriental splendour to many a New York and Washington grand mansion. He also designed and made stained glass windows for his father's friends and for important public

The hand making of a Tiffany lampshade in Christopher Wray's factory. Here Tiffany's own techniques are used to produce opalescent glass in a great many subtle colours and designs. It is an exacting job and the shades, being made of glass and lead, are heavy.

Two mushroom lamps in the Tiffany manner, showing the spectacular effect you can get by using different shapes (one is a squat and comfortable 'toadstool', the other a taller, more elegant 'mushroom') and different colours. The blue is a cooler, more restrained light; the rusty one friendlier.

buildings. After he married his second wife, who was the daughter of a Presbyterian minister, he became more and more involved with church windows and interiors. In fact windows designed by Tiffany were installed in churches in forty states in the USA as well as in Colombia, Canada, Australia, England, Scotland and France. Sadly, fewer than half of these still exist since many of the churches have been demolished, specially the ones in urban areas.

The glass he used was all hand made in different thicknesses and densities, and it didn't take long for him to want to experiment with the use of glass in artificial lighting. His first venture into this was the interior decoration of New York's Lyceum Theatre in 1885. In fact, Edison installed the first electric footlight ever used for a stage there, and Tiffany designed and installed the sconces, which were described by a critic in *New York Morning Journal* as being 'like fire in monster emeralds'. In 1889 Tiffany went to Paris for the Universal Exposition where he was captivated by the irridescent surfaces on glass exhibits, which he was sure he could achieve on blown glass and sell to the public. He also thought he could equal the success of his rival John la Farge who was exhibiting and selling in Europe. Eleven years later he sent two large stained glass windows to the 1900 Exposition Universal, one designed by his staff artist Frederick Wilson and the second, 'Four Seasons', designed by himself.

By 1896 he was making and selling the first portable Tiffany lamps. He took the traditional kerosene burning 'student' lamps, using ready made parts, and finished them off with his own decorative ideas, providing a surface patina and irridescent glass shades. Next he turned to lamps with blown glass bases and shades. He acquired a bronze foundry so as not to have to rely on other manufacturers for bases and in 1898 he produced the first lamp to combine a metal standard with a leaded glass shade, using the techniques of his stained glass windows. People were enchanted and during the next ten years he produced hundreds of similar shades

based on a variety of flowers and plants, all hand made, hand coloured and hand assembled. The price list of Tiffany Studios in 1906 itemises over 400 different models of oil and electric lamps and hanging shades, as well as over 150 candlesticks and over 300 other home and office accessories, not to mention the blown glass products. By 1905 he had nearly 200 craftsmen working for him representing every relevant technique. He had glassmakers, stone setters, silversmiths, embroiderers and weavers, casemakers and carvers, gilders, jewellers, cabinet makers, all giving shape to the carefully planned concepts of a group of artists including himself, who kept a personal eye on everything that came out of the Studios. You will find peonies, wistaria, poinsettia, rambling roses and woodbine in most of his work. But he also played with geometric shapes using turtleback tiles or glass jewels as a decorative fillip. His lamps were in the forms of nature on a grand scale. Sometimes there would be sixteen lilies at the top of sixteen stems springing from a base of leaves. Imagination ran riot; there were toadstools, snails with glowing shells, tulips, lilies, female figures extravagantly and transparently draped. Alabaster, pewter, beaten copper and acid etching were all used. The coloured glass shades were made of minutest pieces of glass, all set by hand and giving an extraordinary feeling of delicacy and richness. Some had prisms which created myriads of little rainbows, some had the most intricate bronze holders for the bulbs.

Tiffany was not the only Art Nouveau lamp maker of course. There were the French, the Viennese, the Belgians and some Scotsmen who were producing brilliant lamps as well. Tiffany's work was, in fact, considered rather vulgar by some people at the time, though that didn't stop it from selling. Nor did it stop people from copying his work, though copyists never managed to get quite the same effect. He was absolutely fascinated by coloured glass and disliked 'perfect' and 'refined' glass. His great admiration was for the medieval glass he saw in church windows and in his search for similar colours he 'took up chemistry'; built his own furnaces and little by little, year by year, began to produce the sort of colours and lustres which could match the old glass in the windows of the twelfth and thirteenth centuries, which had always seemed to him the finest. He found contemporary commercial glass to be totally inadequate for what he wanted and began to research into creating his own glass, looking for

Opposite
A good example of the possibilities of using the flower stem to lead the electric flex up through the foliage to the flowering light at the end. The bluebell has a bud lit by a tiny bulb; the larger flowers are lit by miniature round bulbs. They add mysteriousness to the house plants, but little light to the room.

irridescence, mottling and incandescence. He produced cypriote, fractured and lava glass, but is best known for his favrile shades. (Favrile was the term used for Tiffany's technique for introducing colour into the glass by hand.)

In spite of this tremendous output, Tiffany always insisted on the highest quality, regardless of cost, and used only the finest bronze for his lamp bases. His glass shades were assembled with copper foil and soldered bronze rather than the traditional leading, because that made them more durable. Every part was hand finished; there were no short cuts, and no economies. His were undoubtedly among the best and most expensive lamps ever made. He would use industrial pieces of glass in a lattice of soldered copper foil so that the lamp resembled cloisonné enamel when the lamp was off and plique-a-jour when it was on. Not all Tiffany's lamps were electric. In the early years while electric lighting was running parallel to oil lamps he made many fine oil lamps with favrile glass shades. But oil gives a fainter light than electricity so the shades lose some of their brilliance, and oil lamps have to be constantly cleaned in order to shine their brightest.

As chairman of the company Tiffany retained the right of approval throughout his career, but after he retired in 1919 the work began to lose some of its character and Tiffany Studios was liquidated in 1938, five years after his death. Antique lamps from the Tiffany Studios have become collectors' items and a table lamp with a bronze base and glowing glass shades might now fetch anything from £1,000 to £3,000 and even copies can be worth a great deal of money.

Opposite
The entrance portico of a large Georgian house in England has a traditional comfortably stuffed chintz covered sofa and plenty of pelargoniums. A large space like this requires a proud looking lamp and this large, glowing, rose glass shade on the table lamp behind the sofa fits in very well.

Art Deco

Art Nouveau is often confused with Art Deco but, in fact, though close in time, they are poles apart in thinking. The 1914–18 war really put a stop to Art Nouveau. After the war that freedom of thought and movement was seen by many people to be simply a self indulgence and reaction set in. 'These figures with their tormented and crushed draperies do not seem to me to be really naked, and their proportions are disconcerting – and then the strange idea of adapting an electric bulb as a woman's stomach . . .' wrote art critic Maurice Hamel in the *Revue des Arts Decoratifs* in 1901. By 1918 most people felt the same.

Art Nouveau was the use of pure, undisguised and unapologetic decoration in the home; art for art's sake, in a domestic setting. Art Deco was completely the opposite. The great designer catch phrase of the time was 'Form must follow function'. It was a rigorous, taut, uncompromising style which had no truck with the unnecessary, the flippant or the second rate. Cubism was in and sentimentality was out. Art Deco was full of hard new looks, jagged edges, angles, geometric shapes, loud, positive colours, domesticated for people's homes. The shapes came from Braque and Picasso and the colours from Diaghilev and the Russian Ballet. After Art Nouveau's pastels, the colours were really shocking; garish oranges, emerald and jade greens, purple, every variety of crimson and scarlet.

Opposite
This collection of Art Deco lamps gives a very good idea of the angular shapes and warm colours inspired by Mexican, American Indian, Negro and Egyptian cultures. They are clean cut, warm and uncompromising and fit in well with much modern interior design.

Right
Another authentic Art Deco collection which includes a magnificent ceramic fish, a small coffee set and a tiny black butler, discovered in a junk shop, carrying a glass globe. Art Deco requires a purist attitude if it is to be successful. No untidiness and no compromise.

The discovery of Tutankhamun's tomb in 1922 led to an Egyptian influence on architecture which showed in the Hoover Building just outside London and the Carreras Building in the USA. The Americans were introducing a Mexican and an American Indian influence. It all co-ordinated with negro culture and jazz. The women suddenly became slim and polished with angular hair cuts or permanent waves, cupid bow lips and thin, pencilled eyebrows. It was the time of cinemas, rainbow lighting and chrome. The idea was seeping in that quality and mass production in design were not mutually exclusive, and glass and metal which could be factory made began to take over from wood.

Much Art Deco furniture was well built with the result that it is still possible to find secondhand pieces in good condition which will co-ordinate very well into particular modern environments, although these have rather the quality of collectors' pieces now.

Opposite
In this small corner a modern chrome table lamp base with its cinema-like glass shade is juxtaposed with a motley collection of objects including an armchair upholstered in an Art Deco design. The lamp is made in the same factory and to the same specifications as the original would have been fifty years ago.

Page 40
This is a collector's room. In one corner he had combined an Art Deco sideboard, two original water-colours, a number of ceramic deco figures and artefacts, all lit by the tall deco standard lamp in dark varnished wood with a parchment shade and long tassles. The rich colours are characteristic of Art Deco.

Page 41
Another collection of authentic Art Deco: an angular mirror with a wooden frame and a marble and onyx clock have been lit by two modern deco lamps, both with chrome bases and opaque glass shades. The round shade is on the squared base and vice versa but they could be changed round.

Post War Modern

After the Second World War reaction set in again. Scandinavian simplicity and use of natural materials became the holy grail. Pleated paper lamp shades hanging down or pinned against the ceiling were absolute essentials – much as the ubiquitous Chinese paper globe is today – wooden slatted shades and basketwork shades; all gave a warm, comfortable light and were natural, simple and new. Those who didn't have pleated shades, installed glass globes. They were fitted in bathrooms, halls, staircases, out of doors. New materials were being used including metal and plastic. Spindly metal became fashionable for furniture and was used as lamp bases for both standard and floor lamps, painted black, sometimes with coloured bobbles on the ends. Shades became opaque, plastic, often cone shaped. Metal and glass were also used for the magnificently stylish furniture and lamps which began to come out of Italy. Great slabs of marble formed lamp bases with huge metal arcs spanning several feet with a

Another pleated lampshade sheds its light onto the polished piano top. The glass bowl and pretty arrangement of flowers combine to create a harmonious whole. This is a timeless interior which would have looked good in the fifties and still looks good today.

Previous page
Pleated lamps, popular for their
Scandinavian simplicity and style
in the late fifties, are making a
comeback but in a much lusher,
plusher way, more suited to the
eighties. Here, pleated shades are
used to echo the pleated fabric on
the wall and the pleated and
ruched curtains.

Opposite
The now classic Titzio lamp from
Italy is cleverly counterbalanced
and looks elegant in any position.
It uses low voltage electricity and
its transformer is incorporated in
the base. It is used here to
illuminate the high-fi corner of a
living room, which it does with
great efficiency.

Right
A low, very spare Italian-designed
standard lamp which directs its
light very specifically to a
particular spot – in this case to
illuminate the bust on the
mantelpiece – or swivells towards
the chair as a reading light. The
white base goes well with the
white of the fireplace.

glass shade blossoming at the end. The emphasis was on the sculptural quality of these lamps. Many have become classics as pieces of furniture.

Fluorescent lighting, which gave a clear white light became the accepted working light and was used endlessly in chain stores, offices and kitchens. However it got a bad name for domestic lighting because the colours available were harsh, unbecoming and hard to live with.

By the end of the sixties, domestic lighting was still thought of in terms of single light sources, and many homes still relied on a single light bulb hanging from the ceiling, perhaps backed up by a standard or table lamp in the living room. But architects and designers were beginning to see the potential of whole house lighting 'schemes' using recessed ceiling lights and spot lights. The ceiling light track came onto the domestic market allowing two or three spot lamps to be installed in a cluster which could be angled in different directions and used with different bulbs to get a variety of results, from whole room lighting to lighting particular areas. By the late seventies interior designers were setting up purely to specialise in 'lighting design', largely for offices. Design of light fittings had taken its place among industrial products as a natural partner to furniture design.

During this time tremendous technical advances were being made in lighting wiring and 'systems', and in the development of low voltage lighting together with a better understanding of the quantity and quality of light that would be best for various tasks and lighting functions around the house. Low voltage bulbs cost less to run but the idea has taken some time to catch on domestically partly because of the lack of fittings to be found in the shops. Italian designers cottoned on very quickly to the potential of the tiny low voltage bulb which they began to use in table lamps and lighting schemes. These lamps give a clear bright light, pleasant to live with, good to work by and cheap to run. Advances made in the name of contract designing sooner or later find their way into our homes, and low voltage lights and the idea of actually planning the lighting for a home, is open to all. Because of their neatness and small size, they offer the opportunity for very subtle and efficient schemes of concealed lighting throughout the home. However, one does not want to throw out the pleasure of individual lamps just because we know what modern lighting can do. It would be sad to lose that connection with our ancestors, with their flaming torches and carefully conserved oil. The trick is to combine the new knowledge and to have thoroughly efficient lighting throughout the home while retaining some individuality in the choice of fittings. Many people living in older houses take great pleasure in using lamps which would have been fashionable at the time the house was built. Luckily there is a great deal to draw from, including the antique lamps dating back to the sixteenth and seventeenth centuries (rare now) and antique candlesticks of all kinds, as well as reproduction Victorian lamps, from paraffin and gas to electric lamps in traditional shapes. Then there are the modern, sculptural and imaginative lamps of today's young designers who see that they are not restricted to shape because of the need to store fuel or because of tradition.

LIGHTING
ROUND THE HOUSE

There is now a wealth of choice of styles, as well as a knowledge
of the type of light necessary for different activities: clear
light for working, gentle light for relaxing, directed beams for
highlighting and shadowless light for safety. In this section
are some practical examples of using light efficiently and
decoratively.

Basic techniques

When it comes down to planning lights for your own home, it helps to understand how you can achieve different effects for different situations. The direction and level of light in any situation is governed by two things: the type of bulb and fitting; where it is installed, how angled and whether fixed or adjustable. There is a guide to light bulbs on page 132. The following is a brief guide to the terms used for different ways of angling and beaming light to achieve the desired effect.

Downlighters

These direct light straight downwards and are often fixed to the ceiling or recessed into it. They give good, general light and placed correctly, efficient work light. Several low watt lights can be restful to the eyes while giving quite a high level of light, whereas the same amount of light from one high watt bulb could be less efficient and more of a strain on the eyes. Downlighters are often 'built in' in this way , but of course even a spot or angled lamp can be used to shine light downwards if you direct the lamp that way. Ceramic lamps with fabric shades are largely downlighters with a bit of light escaping at the top, though in this case they are often used as 'pools of light' rather than as task lights.

Recessed fittings

Recessed lights are nearly always fixed in the ceiling and act as downlighters. They are attached by special fixings so as to shine through a hole cut for them in the ceiling. They may be difficult to fix to old lath and plaster ceilings without removing large areas of laths and replacing them with plasterboard. Certain recessed lights generate a lot of heat under the floorboards, which is why it is important to have ample space where the air can keep them cool. They need about 10 cm (4 ins) depth for low voltage or 30 cm (12 ins) for traditional incandescent bulbs. If there is not enough depth you could use eyeball or semi-recessed fittings instead. Use them for general lighting or task lighting in kitchen or bathroom.

Uplighters

These are lamps which direct light upwards. You can simply create a pool of light on the ceiling or light up an interesting feature of the room, a cornice, say. Reflected light from the ceiling can add quite considerably to the level of light in the room, and to the light actually emanating from the lamp.

There are many elegantly moulded and shaped wall fittings available which can be used

Opposite
One carefully planned ceiling recessed fitting is all that is needed to light the sink unit, looking onto this charming window in a converted mill. Carefully chosen sink and tap fittings and one well placed bowl give colour. Other recessed fittings are used over the worktop and cooker.

Right
This is a good example of how an uplighter can shine directly onto a white ceiling and reflect a lot of light back into the room. Although the piano player would need a task light directed onto the music and keyboard, the one standing lamp gives perfectly adequate background light for the whole room.

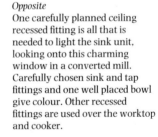

as uplighters. Placed thoughtlessly, as they often are, they are inclined to light up inept plaster or peeling paint. But well placed they can give a decorative and imaginative warmth to a room. Some standard lamps make excellent uplighters. There are many inventive and stylish modern ones which send the light up from a reflective bowl placed on a slim leg. Another form of uplighter is a floor or table lamp in a drum shape which can be used behind plants to show the outline of the leaves, or under and behind stained glass objects so as to light them from behind. Such lighting can be a cheap and easy way to soften what might otherwise be rather bleak general lighting coming from a central source.

Wallwashers

Wallwashers could be called 'sidelighters'. They are wide beam lights used to wash a complete wall or a large part of one with light. Obviously they look most effective where there is something to show off; a large painting, a hanging rug or textile, sumptuous curtains or wallpaper, or an architectural feature, perhaps an arch or piece of moulding. Wallwashers have a separate, rather large curved reflector. A spot lamp angled at an opposite wall is not a wallwasher because it won't give that wide, generous area of light. You can often tell such a fitting by its pudding bowl shaped reflector, though there are some much neater, triangular or box shaped ones which also use low voltage bulbs.

Opposite
Another example of how an uplighter, by throwing its light away from an object and onto the ceiling can, indirectly by its reflected light, illuminate that object very efficiently. The ceiling must be in good condition and should be white or a very pale colour if the light is to reflect off it.

Below
This miniature light on a low voltage track is manufactured by Concord Lighting. It has optional flaps (for use with flood only) to control the beam. Standard track is also widely available.

Task lighting

This is the lighting which enables you to do 'tasks' involving concentrated eye work and for which you need a clear, unobstructed and precise view. Such lighting should bring out true colours and not cause any reflections or glare to tire the eyes. In large offices, task lighting is often provided by fluorescent lights overhead, supplemented with personal angled lamps on individual desks, since the ability to angle a lamp is important to task lighting so that a person can switch from one activity (say word processing or machine sewing) to another (say, answering the telephone or sorting colour swatches) and bring the light to bear on each activity at will. Adjustable task lamps can be desk, floor or wall mounted, or can clamp or clip onto a shelf, desk or worktop edge.

Pendants

Pendant is the generic term for any lamp hanging from the ceiling, from a bare light bulb to an intricate chandelier. In the main they provide general lighting rather than lights for specific tasks, though a pendant over a table, or a row of pendants over a worktop, count as downlighters and can be satisfactory as task lighting if placed correctly. 'Rise and fall' pendants on a pulley-system are available, which are useful if fixed over a kitchen/dining table or a dining/work table so that the light can be heightened or lowered as required. Pendants can be looped from a central ceiling fitting to a not-so-central position. They may also be hung with a long flex, quite low over a coffee or other occasional table.

Opposite
Traditional brass shell lamps on a single base are highly flexible. They can be twisted to face up, down or sideways. One can be directed towards a book or writing pad, the other upwards or onto some other work area. They are positively suitable for large desks where their interesting shape can be enjoyed.

Below
A chandelier pendant can give several sources of light where a single pendant gives only one. Such lights are also much more decorative than single lamp pendants, giving more light at the same time. This one has five sockets for bulbs, each shaded by frilly, glass, Victorian gas type shades.

Fluorescent tubes

There is a coating of phosphor on the inside of a fluorescent tube which is activated by ultra violet radiation within the tube, making the phosphors glow brightly and evenly. Fluorescent light uses something like five times less electricity than incandescent to give the same amount of light. Fluorescent tubes also last longer and radiate less heat. Recently fluorescents have been produced which offer much more comfortable colour rendering so there are good reasons why they should be installed as task lighting in certain parts of the home.

All fluorescents require special control gear, including ballast (or choke) to provide the initial surge of electricity to start the tube up and thereafter to control the amount of electricity consumed once the light has 'caught'. You can get rapid-start tubes which are slightly more expensive. Modern fluorescent tubes are not messy and shouldn't leak. The colours have been much improved by the use of different phosphors and new designs for the fittings available for fluorescent tubes have improved very much as well.

Opposite
A ceramic bowl lamp of the kind you can buy everywhere has been given a remarkable and dramatic quality simply by framing it in a window and putting a china cat on either side. The combination of dark background and glowing yellow shade is quite stunning.

Below
Purpose designed picture lights are quite unobtrusive and can be fixed above paintings, prints or small wall hangings to illuminate the image. In this case the light also catches the several objects grouped on top of the mantelpiece in a very effective way.

A deceptively simple little Italian lamp uses a low voltage bulb and has a lid that lifts up or down to direct the light. It has been used on this tiny table to shine directly into the glass bowl, making it glow with a wonderful clear light.

Above
The rather harsh 'whiteness' of fluorescent is lost completely if the tubes are concealed behind a baffle, or right at the back underneath cupboards above the worktop as in this kitchen. The light reaching the worktop is efficient and gives a warm, friendly glow.

Below
This double bracket is not giving off a tremendous amount of light behind its yellow globes, but the light is cheerful and sunny. It helps to add colour and interest to a wall which would otherwise be a bit dull and tedious. The light is adequate as general background light, for safe walking about the house.

Neon
Neon, or more correctly, cold cathode lighting is produced by tubes containing neon, argon or krypton gases. These, together with phosphors, like the ones used to coat fluorescent tubes, can produce twenty-five or more colours. Such fittings operate on a very high voltage and require transformers with special kinds of cable. But neon is becoming popular for special effects lighting at home and safer fittings using low voltage input are being developed. All the same, these lights use a high voltage, are vulnerable to breakage and should not be used in children's rooms. They are expensive to buy, though once installed, neon lights are economical to run and

should go on for ever. The ability to bend the tubes into more or less any shape is the reason they were so popular in street and other signs from the twenties onwards, and why they offer opportunities for superb flights of fancy. The high pitched buzz produced by the transformer could be irritating and they need to be put somewhere out of harm's way because of their fragility.

Pools of light
Such lighting normally serves no purpose except to make a room more interesting, softer and friendlier, though occasionally the light may be at a high enough level to read by, as

when it is supplied by a table lamp with a shade which acts as a downlighter as well as a diffuser.

Wall sconces can act as 'pools' as well as uplighters. And lamps such as the light-up farmyard animals, many 'Tiffany' and other glass lamps, glowing fittings in glass, plastic or other materials or fittings which are not themselves seen but which shine gently behind plants, all provide pools of light which help to soften and warm an environment.

Highlighting or accent lighting
This is the lighting used to bring to life a painting or a plant, a sculpture or a print, a dining table or a coffee table or a particular feature of a room you want to make the most of. It is quite difficult to achieve the right effect because there is such a wide choice of beams and angles. Try to make the beam fit the object. Too wide a beam will cause the object to get lost; too narrow a beam will only catch a detail so that you lose the basic shape. Find a position where the beam won't hit you in the eye as you walk past or under it. Before you finally fix the lamp, check the beam angle and make sure that the light can be adjusted correctly. Such effects are most often achieved by spotlamps or low voltage beams. But a low pendant can highlight a small table and its contents with great effect and if carefully chosen, the fitting itself will contribute to the general 'look'.

Low voltage lighting
Most low voltage bulbs consist of minute tungsten-halogen capsules integrally set into a reflector and this reflector determines the angle of the beam. Low voltage bulbs use the light very efficiently. Because of this, one low voltage bulb can take the place of a mains voltage one of twice its wattage. Low voltage bulbs also last for 2000 to 4000 hours – more than twice the life expectancy of a conventional GLS bulb. They are also very small, which has enabled designers to come up with many small, neat fittings both for built-in and individual table or floor lamps, and for ceiling fixtures, where they can take the place of more cumbersome spotlamps. Just as fluorescent lights require special control gear, so do low voltage bulbs require transformers.

Low voltage lighting is cheap, effective and much more available than a few years ago although, unless just buying a single lamp with a built-in transformer, you will have to have it installed by a professional. In many low voltage lamps and fittings the transformer is an integral part of the design.

Halls, staircases and passageways

Every home should be warm and welcoming from the outside. It doesn't matter whether it's an apartment in a large block, a terraced house, a loft conversion reached by an old service elevator or a cottage in the country, the first glimpse should be like a beacon in the dark. Even a candle in the window can send a message of welcome to those on their way home.

Entrance halls are not always easy to deal with effectively. They are often forgotten as part of the personality of a home and all the effort is put into making a beautiful living room. However, first impressions are all important and family and visitors should all be made comfortable and at ease almost before they set foot in your home. Hall, stairs, passages and corridors are all really part of the same thing and should

Opposite
A long and narrow hall has been given a real feeling of spaciousness by the mirror tiles completely covering one wall. Because it is very high, it has been possible to install three spot lamps and the wall bracket is reflected in the mirror, greatly increasing the amount of light received from it.

Right
A small corner of a hall in a Victorian terraced house in London has been papered with a tiny floral print. The miniature painting is very much in keeping with the print and so is the old fashioned gas type glass lampshade which is faintly tinted blue to match the paper.

be treated in similar manner to give the place continuity. In the first place, take some care with the decorating so that any lamps will actually be illuminating something worth seeing. Pleasant paper, carefully chosen paint, posters or paintings on the wall all help to give personality to this most important of spaces. Small, narrow halls, as so many are, will benefit from mirrors – the larger the better – which seem to enlarge the space and lighten it too. Narrow shelves and elegant hooks for hats and coats make the place more practical as well as more interesting. Plan your lighting in conjunction with what's in the hall, and install any extra wiring before you do the decorating so that it can be channeled into the wall.

You need good general light, not just to make things more cheerful but also to see notes and letters and peer into the darker corners for wellingtons and lost toys. Many halls have only one pendant light hanging from the ceiling. You can make the most of this by giving it a high wattage bulb with a spectacular shade. This may also be the area where you can give reign to your secret yearning to own a chandelier. The central 'stalk' with its several arms holding smaller wattage candle bulbs may provide a more interesting and effective light than the traditional single bulb. Alternatively fix lighting track instead of the pendant and angle two or three spotlamps so that you get a variety of effects and a stronger concentration of light.

Opposite
In this square, quite spacious hall, the owner has room for a pedestal (made of polystyrene) to support a black lamp. The black and white of the shade is echoed by the floor tiles, frieze, black framed portrait and black table with its black vase of artificial black tulips.

Right
This converted paraffin lamp is fixed to the wall of the stairs by an ornate metal bracket. It gives adequate light for safety and adds character to a part of the house which is often difficult to deal with. The light is diffused through the white shade which helps to give as much illumination as possible.

You may like to add a wall fitting or two to this. Unless you have a spacious hall it's best to have ceiling and wall fittings rather than anything standing on the floor, simply because the tidiest halls already have enough paraphernalia in them. Small halls should be free of floor-standing equipment, if only so you can vacuum easily since any hall opening straight out onto the garden or the street inevitably gets much dirtier, much faster, than other parts of the house and you need to be able to clean it without bother.

More spacious halls offer greater possibilities because you can stand an attractive large lamp on a hall table and allow this to be a feature in its own right. Even the bleakest of lobbies often has interesting architectural aspects such as an arch, or covings, or elaborate cornices, or staircase structures which would benefit from

an uplighter. There may even be some stained glass lurking about which will be much enhanced if light can be shone behind it. If there's no built-in feature worth highlighting, maybe a kelim or Persian rug, or a modern tufted rug could be laid on the floor with a downlight shining onto it from the ceiling. Provided it's a good thick one and not liable to skid from underfoot, rugs are an excellent way of bringing the eye to something bright and pleasant, and away from something perhaps more tatty and uninviting.

From the hall or lobby it's necessary to consider the inevitable staircase. Staircases are not easy to light and you must be careful that ceiling mounted fittings on stairs don't shine into the eyes of the people coming in at the front door or about to climb the stairs. Spotlights can offer perfectly adequate lighting provided they

Opposite
A metal Art Nouveau type bracket in the shape of a woman reaches out from the wall to lighten a hall, stairway or ante room. This one looks well with the heavy curtains, rich wallpaper and the small table with its exotic dried artichokes in a large decorated vase. A nicely opulent treatment for the right space.

Right
This entertaining glass shade is made in the shape of a pineapple, but its colour is more exotic than any pineapple from real life. Used here, next to the rather spiky plant in a pot, it gives a foreign, jungly effect to what would otherwise be a rather mundane corner of a room.

can be adjusted to shine onto the wall beside the stairs rather than directly downwards. But other types of wall lighting can also be very satisfactory. Wall uplighters are attractive and, provided the light level is enough to make the treads and risers clearly visible without shadows, should work very well. As with other parts of the home, it is better to have several sources of low level light than one brightly shining bulb. The single bulb is much more likely to cast confusing shadows – very dangerous on stairs. And although, of course, the lighting must be adequate to see clearly where you are going, it doesn't need to be at as high a level as, say in the kitchen or living room, because it is not being used for close work, but simply to guide the feet. Safety considerations say that the lighting you use should not cast shadows over stair treads. The edge of each tread and the depth of the whole step should be clearly seen. You can combine a general diffused light, say a frosted glass globe or a Chinese pendant lampshade, with the softer light from wall sconces up the stairs. Place such lights carefully, experimenting and testing before fixing them for good. Every stairwell is different and the length and width of the staircase, the height of the ceiling, the depth of the treads and the height of the banisters will all affect where you should place your lighting.

One modern and effective stair treatment is to have low wattage bulbs on every step (or every third or fourth step) to shine onto the treads. These must be combined with general light from above and you will have to devise a false wall to hide the wiring behind. If you have a wooden stair rail against a wall, you can use this as a baffle and put strip lighting underneath it. This can be either incandescent 'architectural' tubes or fluorescent tubes. It will give a pleasant soft wall washing light, very adequate, safe and good looking at the same time.

A two way switch is sensible if you have more than one flight of stairs. Dimmer switches allow you to leave the light on at night in case people wake up and like to walk around.

Corridors and passageways are hard to light but the lighting is very important. The longer the passage, the more light fittings you will need. Somehow, you have to get across the feeling that this is not a long walk into gloom but a cheerful space. The wider you can make it seem the less bleak it will be. Mirrors can be highly effective in this way, and the use of pale colours rather than dark ones. A row of wall uplighters will probably be more effective than trying to deal with the whole space by using a spot or two. Paintings along the wall, each with its own light will also help to bring such an area to life.

Opposite
A narrow or otherwise bleak hall can be much cheered up if you can find room for some form of decoration to make it look as though someone cared. Here, a narrow chest holds a basket of dried flowers and above it a painting with a fine wide open landscape is lit by a picture light.

Living rooms

Most living rooms these days have to perform several different functions. Few people can afford a separate 'parlour' or 'front room' which is only used for sitting and chatting to visitors. The living room is certainly used for conversation, but also for reading, watching television, writing, playing games, giving parties, hanging paintings, showing off objects and very often for eating as well. So living room lighting has to be very carefully thought out. The important thing is to have a flexible lighting system which includes several different kinds of light so that they can all be turned on at one time or which you can orchestrate, depending on the atmosphere you want to create and the tasks you want to do.

Many homes are still blessed (or rather, cursed) with the very first form of domestic electric lighting, when each room was supplied with a single light fitting hanging from the centre of the room. Many houses built today still have this central fitting and not much more, but the concept is very out of date because such lighting is hard, inefficient and boring. However, this doesn't mean that hanging lamps have to be done away with completely. It's perfectly possible to extend the lead, loop it and hang it over a table which is not in the centre of the room. For a work or dining table, it should hang a little above the eyes of people sitting at it, so there's no glare; for a low coffee table it can hang right down over the books or magazines

One lamp may be used in several ways. Here two modern Italian uplighters have been fixed to a large wall to give general background illumination to a seating area with a large and comfortable settee. The light shows off the moulding and reflects off ceiling and walls to give a pleasant, calm atmosphere.

or whatever you keep on the table. If the bulb is visible from below you should use a crown silvered one, or at least an opaque or pearl one to prevent glare. Shades available are many and various; from traditional to high tech, brightly-painted metal ones of the sort used in ancient factories, to the ubiquitous Chinese paper globe or the more recent Japanese squared paper rectangle, or even ones like Victorian mob caps. Burnished copper has a warm sheen.

If you prefer something more like a chandelier with several bulbs sitting upright, there are plenty of those too. Here, traditional designs come into their own. And there are many gas type lamps on the market with the prettiest glass shades, either plain or frilly available. If you want to go the whole hog and get a chandelier with glass which will reflect the light, be sure you have an environment grand enough to do it justice. True chandeliers can look very out of place in the wrong room. They need a carefully thought out plan which will match their opulence. Low level general lighting from many sources gives a much pleasanter environment than bright, harsh light from only one or two sources and can easily be supplemented as and when necessary.

The living room, of all rooms, is the one where you can match the lighting to your interior decoration, and indeed, to your mood. Owners of historic houses often enjoy lighting them as nearly as possible in the style of the time

In this small alcove, the same fitting is used nearer to the objects beneath it, and the reflected light is at a high enough level to enable its owner to see the hi-fi system perfectly clearly, so that in effect it has become a task light, though of a very gentle kind.

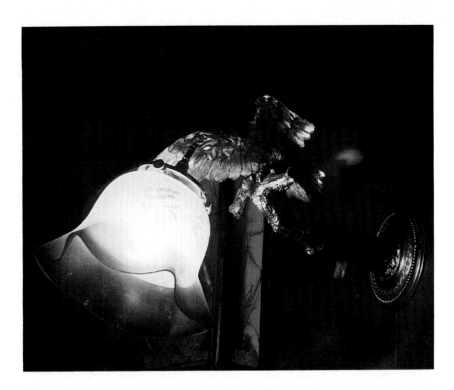

they were built. Many such home owners do still use traditional pricket candlesticks, wall candlesticks with silver reflector panels and floor standing iron candlesticks, with no overhead lights at all. Some of the candle holders were actually made at the same time as the house, not just in the style thereof. Tiny low voltage spotlamps can be hidden discreetly behind objects on old chests to light any interesting collection. Many National Trust houses in Britain are painstakingly lit in this manner and the people living in them are careful to preserve the atmosphere and look of the house as it would have been. Although comparatively few people have medieval, or even very old houses, many do admire and enjoy the softness and warmth of such lighting and can use a similar sort of lighting to get the general effect. The candles often used on dinner tables are an example. But if you want to take it further, there are plenty of reproduction lamps around which are very similar, and in some cases actually made in the same factories and with the same machines as the originals.

Let us suppose that you already have some form of general lighting; a pendant ceiling lamp or recessed downlighters. You are then free to consider how to tackle the more subtle and individual requirements of your life-style and decor. In a living room you want to create a feeling of relaxation and warmth. The lighting should always be subtle and the light source

Opposite
A pretty living room which shows the advantages of flexible lighting from various sources. Each lamp does a specific job; the wall brackets giving general light and the table lamps acting as pools of light, and in one case as display lighting as well.

Above
This is one of several patterns of glass shade on the market. They are based on the traditional Victorian gas shades and are available with a variety of wall brackets. They are particularly suited to the sort of flowery, pretty interior which goes with Victorian town houses and certain cottages and should not be used in modern or Art Deco interiors.

In this modern room, designed by an architect with Scandinavian leanings, the lighting has been carefully designed to be simple but flexible. In the first picture (*opposite*) the two tracks of spot lamps, angled outwards to light up the objects at the sides of the room, bring the whole room into focus, showing up the polished wooden floor, the sculptural furniture and the books. In the second picture (*left*) the spot lamps are switched off and only the copper pendant is lit, leaving the main part of the room in obscurity and focussing on the work on the table. The same effect could be used for dining.

should never be too obvious. In a room primarily designed for leisure activities, you can afford to put in lights simply for their own sake as well as those which do a particular job. Pools of light whose function is simply to please and intrigue are eminently suitable for the living room and provide constant interest. They are soothing to the eye, just as soft music is soothing to the ear. Ceramic vases or bowls are effective, providing diffused light through opaque shades and small areas of downlighting as far as the table, or floor, they stand on. There are many, many variations of the ceramic bowl from enormous painted Chinese jars, Japanese vases, Portuguese animal shapes or simple round balls. The ceramic part is highlit by its own beam and the shade can be chosen to match the decor or to contrast with it. The shade diffuses the light and disseminates it. The choice of colour will make a great difference to the final effect. Dark shades will allow less light through than pale ones and the shades can be used to add a touch of colour during daytime. Most such fittings

This rather elegant but not specially unusual ceramic lamp and shade demonstrates perfectly how an ordinary lamp can be used to create a 'special effect'. It's placing on a small table in front of a curtain of a particular colour and the choice of ornaments and flowers in similar tones of orange and red make a little stage set.

Opposite
Two flower shaped lamps on narrow stalks give a gentle background interest and enough light to see the painting and even to read by. The difference in height of the two lamps makes the light more flexible and efficient and the flowers themselves more realistic.

Left
The ubiquitous small ceramic bowl can take on a different character depending on the shade used and where it is placed. This one uses a dark shade so that all the light is directed downwards to light the photographs below; very little is diffused through the shade.

Overleaf
Because this is a large room, the lighting has to work rather hard to be effective. Pale walls and pale lampshades help to make the most of the light there is. The uplighter reflects light off the pale ceiling and as there is no central light, the overall effect is calm and pleasant.

take standard light bulbs and the light coming through will be yellow, which will affect the colour of the lamp when it is on. Such shades can be pleated, patterned, parchment or fabric; deep and narrow or wide and shallow, but they must always be large enough for the base. Nothing looks worse than a good sturdy base with a pathetic little half-hearted shade perched on top. Large lamps of this kind can sit on the floor, smaller ones on small tables, window sills or on a television set (where it will not cast its reflection onto the screen).

Besides these very simple lights, there are the traditional oil and paraffin type lamps of the kind invented by Leonardo da Vinci, working into the night in fifteenth century Italy, and by many writers and draughtsmen at their desks until the beginning of this century. These have pleasing proportions and many are just like the originals. You can, of course, still get lamps which will burn oil and the slightly weaker light can be very attractive, though such lamps do require constant cleaning and a modern household would probably only want to use them

under special circumstances. The modern, electric versions can look very charming and really do a better job than the old ones. Traditional oil type lamps are often made in brass with a choice of glass shades. They need space and height if you want to display them to their full advantage, and do look their best when used with furniture of the same period; that is to say with solid wood, old fashioned desks and bureaus, gate legged tables, military chests and so on.

Still on the subject of background lighting and looking at 'pools of light'; wall sconces are popular because they provide sculptural interest to a plain wall with a gentle light emanating upwards from them. These are not used for practical purposes to read or work by, but give a very comfortable, cosy glow, and can look very effective if used under a moulded ceiling or frieze. Sconces may be glass, ceramic or metal in a great variety of styles from Art Deco to high tech modern; in simple ceramic triangle, bowl or shell shapes to stained glass confections or wide metal dishes. There are also wall double lights with little shades, reminiscent of the sixties if they have smoked glass shades, or of the Victorians if they have little framed fabric ones. Other wall lamps can imitate the old gas lamps of yore, with frilly, etched, plain or coloured glass shades. These give a very pretty light for a 'pretty' environment, complementing exactly that floral, English or Colonial look where modern metal or thirties stained glass would look quite wrong.

Look critically at the room before choosing: a stained glass sunburst will go ill with a chintz and small print 'English Rose' look, but beautifully with squared-off shapes of an Art Deco interior. Simple bowls will go well in many styles of room, where their simplicity won't clash with the rest of the decor. You must, of course, make sure the wall and ceiling which is lit is in good condition, because the light shows up every bump in the wallpaper and every flaw in the plastering or paintwork. What it will not do, is light any picture or object which is underneath it. If such an object is well lit in its own right, or the shape of the light fitting echoes the shape of the image beneath it, the two can form an interesting duo, but as far as lighting is concerned, it will be doing nothing.

There are a number of modern 'designer' lamps intended specifically to provide small areas of light in an interesting form. For instance;

Left
Two paraffin type lamps adapted for electricity. Each makes an excellent secondary source of light for the living room and can be used as an object in its own right; as a reading lamp if placed on a work height table or as a pool of light among foliage plants.

Right
Another traditional paraffin lamp brought up-to-date and converted to electricity. This one uses a really deep sea green to draw the eye towards a group of objects which rely on form rather than colour to make them interesting. The lamp rises out of this group to command the whole arrangement.

Page 80
Because it is small a stained glass
'setting sun' can hold its own
without dominating above an old
fashioned upright piano where no
particular style predominates but
where colour, size and shape are
the guiding lights. It blends in
with the Victorian biscuit tin, the
living plant, the dried flowers and
the modern stool.

Page 81
A well directed single spot is
enough to illuminate this green
table with its fascinating
collection of objects; including a
carved frog from the Phillipines
and a medieval candlestick. All
lights are reflected in the mirror,
heightening their effect. The tall
plant, just outside the main beam,
gives height to the whole.

Right
An imaginative and highly
decorative way to use a modern
Italian standard lamp, to hold
together two narrower sets of
shelves on either side of the
alcove. The complete
arrangement now appears like a
sculpture or a painting with the
carefully arranged pottery
adorning the shelves.

the kite light, made of parachute nylon, and fixed to the wall; the many sculptural, angled floor or table lamps designed in Italy, as well as opaque glass 'vases' and so on. There are many around.

Another form of lighting ideal for living rooms is the standard lamp. Standard lamps are versatile in that they may be used as uplighters, as reading lamps, behind armchairs or as downlighters, or simply as pools of light. Be warned against short standards with two adjustable spots. These can be useful but there are many badly designed, badly made lamps of this kind which are often top heavy and liable to topple over. Check that it will take a reasonable high wattage bulb, that you will be able to get the bulb in and out again when it needs changing and that the fitting will stand up to

being pushed up and down on its stand so that you really can angle it as you wish. The old fashioned standard lamp used to have a wooden pedestal, more or less ornate according to taste, with a simple parchment or fabric shade. Such a lamp placed behind an armchair could be useful as a reading light as well as adding a little interest to a room whose main light was the traditional central ceiling pendant. Another form of standard lamp consisted of a large sculptural reflecting bowl on a stand which would throw a marvellous glow upwards. Nowadays there is a very large choice of standard lamp and modern designs are spectacularly inventive and decorative. Many use low voltage bulbs to give a pleasant and flexible source of light which can be adjusted, angled or simply act as uplighters. These elegant lamps

A modern standard lamp, because of its opaque, light coloured shade, contributes to the lighting of the whole of this corner of the room, but also acts as a reading lamp for anyone sitting in the armchair. The elegant spareness of the design looks good with the varied but carefully thought out setting.

really need an elegant and very sparse background if they are to have their full impact.

Another knotty lighting problem for many living rooms is how to tackle the fireplace. Central heating threatened at one time to oust the traditional fireplace and in apartment blocks fireplaces are not often seen, but in country areas or traditional houses a fireplace does provide the perfect focal point for a room and should be taken advantage of. In winter when the fire is lit, it gives its own quite enchanting light. At other times it can be used as a showcase for objects and a frame and a foil for plants or flower arrangements, and lighting here can add a touch of magic. This is the place for pools of light. The shape of a glowing orb or a stained glass confection can really lift the fireplace and give it a feeling of purpose. Alternatively a tiny spot lamp discreetly placed in front of the grate, can show off the structure, specially if it has pretty tiles or a marble surround, or even an arrangement of plants and objects.

A fireplace is such an obvious centre of attention and yet when there is no fire burning in the grate it is sometimes hard to see how to make the best of it. Lighting can be the answer. The fireplace opposite is being used as 'staging' for a collection of indoor plants and small objects of interest. The corrugated paper light is being used as a foil for the greenery and casts a gentle glow onto the edges of the surround. The smaller fireplace has been treated as a frame for the dried flowers in the grate. The vase-light accentuates the moulding on the fire surround and the pretty pink walls, and silhouettes the true vase with its arrow like contents.

This brass lamp casts its light downwards to catch the sparkle and shine of its own base as well as of the small filigree silver dish and coffee machine. The objects have been carefully chosen so that they reflect the same shiny metal look and solid 'reliable' quality of the lamp itself.

There is one other most important form of lighting which applies to nearly all living rooms and that is the lighting of paintings, prints or any other object you may be proud of, pleased with and enjoy looking at. Unfortunately, most rooms are not equipped with lighting to make the most of such things, though there are several ways to achieve a satisfactory effect. The most commonly used lighting for showing off objects is spotlighting, with one or two spot lamps judiciously placed in the ceiling. Picture lamps are also used which may be fixed to the picture frame to shine down and light up the whole canvas. If you use spotlamps, they should be fixed not too far from the object to be lit. Certainly not at the other side of the room, because even if the ceiling is quite high the light will still glare into the eyes of anyone who gets in the way of the beam. If you are lighting anything precious, remember that light damages textiles, so keep the lighting at a distance and the lighting level low; low voltage lighting is better than incandescent bulbs.

As in all rooms which incorporate many different light fittings, it will be to your advantage to see that a number of them are on the same circuit and can be switched on and off at the same time, from the same switch.

Kitchens

The kitchen offers good opportunities to plan the lighting and install a sophisticated scheme because, once the whole kitchen itself is planned and installed, you are unlikely to change it and lights designed to shine onto work surfaces will not have to be altered later. You can therefore afford to put a bit of money into installing exactly the lighting you require.

Being one of the most complex rooms in the house with many different activities taking place there, the lighting needs careful planning. Most importantly, it is a workshop for the cook, who will need to see very clearly for chopping vegetables, cutting meat, gutting fish and stirring hot pans, all of which need sharp eyes and good light. But as it may possibly be used as an eating room, perhaps a room for entertaining and also for organising family accounts, you will need several different forms of quite specific lighting, most of which will not need to be at such a high level as for the serious cooking jobs. So you will need, not just task lighting for the chopping and stirring, but also the general background lighting which all rooms need, and lighting over the table or breakfast bar for eating or reading (perhaps even for doing homework) and possibly 'highlighting' to show off a dresser with decorative china or some aspect of your kitchen decoration which you particularly like. For such varied needs it is best to have separate switches for each type of lighting; all the task lights could be on one

Excellent task lighting is provided over the worktops, sink units and cooking surfaces of this custom built kitchen. A little extra interest and warmth is added by a fluorescent tube under the central cupboard which, incidentally, adds to the task lighting just there.

switch for instance, and all general lights on another, with perhaps highlighting and table lighting again on separate switches. If you have them all on dimmer switches it gives greater control over the type and level of light and you will find you can alter the whole look and feeling of the kitchen just by altering the lighting.

For a long time it was the fashion to treat the modern kitchen rather like a hospital operating theatre with smooth, shiny 'hygienic' surfaces, fitted cupboards covering every wall and white appliances. In fact, only recently has it become possible to buy electrical goods for the kitchen that are not white. In line with this thinking, the fluorescent tube was considered the ultimate in good lighting for kitchens, giving a clear white light and casting no shadows. Unfortunately, fluorescent is rather bleak to live with and to work by. Even though there are 'warm' colours specially formulated for domestic use such as 'De Luxe Warm White' and 'De Luxe Cool White', which are certainly an improvement on the non de luxe colours, the light is still uncompromising and rather factory or hospital like. Modern thinking for kitchens is to move away from the laboratory look towards something more friendly, less forbidding. You can, after all, be just as hygienic on a green surface and on wood as you can on white and laminate. Over the sink and other work surfaces there should be at least two 100 watt bulbs or two 75 watt reflector floodlights. Many people still like to use one or two fluorescent strips over their worktops, and it is certainly possible to install them so that the light is less harsh by fixing them under wall mounted cupboards, right at the back where you can prevent them from shining right into your eyes. Or if they are fixed towards the front, they should be shielded by a wooden baffle. Objects stored at the back of the worktop are also illuminated in this way and the general result is a gentle and pleasing light in the room. Better still, the tube can be fitted above and at the back of head height cupboards so that the light shines upwards to be reflected off a white ceiling onto the worktop. The high tech, laboratory type of kitchen/dining room will look well if given big enamelled metal shades, copied from those used in factories. These are available in many different colours. Ultra modern kitchens need ultra modern lights, and the ones coming out of Italy are just right. Eclectic kitchens with collections of jugs, china, dried flowers, hanging garlic, old fashioned bits and pieces can have exotic lighting such as Tiffany lamps or stained glass wall

The colours in this glowing Tiffany style glass shade match those of the table cloth in this country cluttered kitchen. The light is fairly low so as to light the whole table without glaring into people's eyes. The rest of the kitchen lights and those in the far room have been dimmed to create an intimate atmosphere.

Overleaf
Here is a combination of task lighting which also provides general lighting and an added touch of the 'intended' with a green glass shade which matches the dark green of the traditional cooking range. This is a small kitchen and the one light over the cooker and a spot light over the table/worktop (not seen) are sufficient to give satisfactory illumination.

sconces, but those are for people who are fairly confident about their tastes.

If you don't care for fluorescent, there's no reason why you shouldn't have perfectly adequate and, many people find, much pleasanter tungsten lighting instead. An efficient alternative to fluorescent tubes can be from spot lamps fixed to the ceiling, or to track on the ceiling or wall. You can get two or three spot lamps onto one track which can be directed downwards onto worktop, sink and cooker. Then there are ceiling lights or recessed ceiling lights placed above the work areas so as to shine directly down onto them. All the above give excellent working light provided they are carefully placed and directed. Traditionally these have used tungsten bulbs but there's no reason why you shouldn't have low wattage lighting if you have it professionally installed.

General lighting

So much for the working lights. If your kitchen is designed to sit in – if you use it for coffee and chatting and enjoying leisure moments, as well as for food preparation – then the whole room will be softened by some gentle general lighting to use either together with or instead of the task lighting.

If you are lucky enough to have a certain amount of wall space not covered by cupboards or shelves, wall mounted fittings can add light in pleasing pools. Make sure you choose fittings that can be easily cleaned since even the most efficiently run kitchen attracts more grease, fly droppings and dust than other rooms in the house. This is a form of kitchen decoration not given much thought normally, but most people don't really need the number of wall cupboards forced on them by kitchen manufacturers and designers. Lighting can completely alter the atmosphere. Wall uplighters which wash the wall with light, or wall brackets with one or several bulbs and shades, are easily fixed, and without those overbearing cupboards, you will have room to hang pictures or prints, which the lighting can help to display. There are lamps to suit every style of kitchen. Country kitchens look well with reproductions of oil and paraffin lamps, with fringed fabric shades. Unless you have a very capacious kitchen it's best to rely on hanging or wall lamps, otherwise they just get in the way and take up food processing space.

Table lighting

If the kitchen is also used as a breakfast bar or has a complete dining area – and perhaps is used for homework, as many are nowadays –

for both space and convenience sake, one of the most satisfactory types of lighting is the rise and fall pendant. (See page 94 Dining Rooms.) It's not necessary to have a rise and fall, of course, you can have a simple pendant lamp over the table, but make sure the light falls over the whole table and that the crown of the bulb is silvered so that the light doesn't dazzle. For a rectangular table you may prefer to have two or more lights hanging down over it.

Highlighting

Many kitchens have decorative displays of china or glass, of copper pots and pans, or other collections. To do them justice, they should be

An old fashioned paraffin chandelier is used here, not converted to electricity. Its gentle light is preferred by its owner for evening dining, rather than brasher electric lights, and its style suits the rustic look of the room. Smoke catchers are used to prevent 'lamp black' from settling all over the ceiling.

well lit. You can use spot lamps quite satisfactorily, perhaps one from the ceiling track you are already using for your task lighting; otherwise from strip lights at the back and top of the shelves or running up the sides of the shelves. These will probably need baffles to conceal the light source, since in this case it's the illumination and not the fitting which is important.

Don't forget the natural light

You may be lucky enough to have a large picture window which looks out onto a garden, a country view or interesting townscape and which lets in plenty of light during the day. Do make best use of these. The light may have to be supplemented on dull days or if the worktops are not in direct line, but natural daylight is a great mood lifter.

If you have windows which look out onto a dismal area, forget about the view and use the light to display a collection of plants or glass, on glass shelves fixed across the window. Coloured glass bottles, stained glass boxes and decorations all look best with the light shining on them from behind. Some plants show up the subtle colours of their foliage quite wonderfully that way too, and of course, it's an ideal little nursery garden, both light and warm, for small plants and cuttings.

Safety

Of all the rooms in the house, the kitchen takes the most planning as far as lighting is concerned because, for safety's sake, you don't want to make up for lack of sufficient light by using portable lamps plugged into wall sockets. Not only will the flexes get in the way and possibly get wet, particularly if the sockets are anywhere near the sink, but the lights themselves will take up valuable shelf or worktop space. Most kitchens in any case don't have enough sockets for all the kitchen appliances, so the rule is no 'occasional' lamps on worktops, tables, clamped to shelves, or even on the floor. In the kitchen, everything should be planned and wired in.

The other thing to watch for is that lamps are positioned so that they don't shine into your eyes, nor cast shadows over any work area. Spot lamps and silvered bulbs are available which minimise glare, but the positioning of the lamps themselves is of first importance, They should not be placed between you and the worktop or your own shadow will prevent you from seeing properly. If you are going to spend money on a complete rewiring and lighting scheme, experiment and test to make sure, when everything's installed, that every light is of the sort you want, exactly where you need it.

Opposite
Simple kitchen units with purpose designed shelves make an interesting kitchen, lifted right out of the ordinary by the lighting. One ceiling spot lights the worktop while a crescent moon and a glass globe help to highlight the mixture of practical and antique objects on the wall.

Dining and eating areas

Eating is one of the basic pleasures of life. It doesn't matter whether you are grabbing a quick bowl of cereal before rushing out to catch a train, sitting down to a family brunch or settling in for two or three hours of formal dinner; to get the most out of your food, eating should be pleasurable and the environment should be relaxing and pleasant. And, of course, lighting can play a large part in how pleasant that environment is.

To own a separate dining room is somewhat of a luxury these days. Most households have to either cook and eat, or live and eat, in the same room. In this case flexibility is all important. You will not want to have to eat your food under the same uncompromising task lights as those you cook by, nor will you want to have to be aware of the cooking debris, necessarily, once you've finished the cooking and can start the eating. So, in kitchens which share the cooking and the eating, it is wise to have two different forms of lighting on two different circuits, one serving the task needs of the room by ceiling recessed downlighters, or other suitable shadowless light (see page 50) and the second by rise and fall or other pendant lamps, over the eating bar or table. You can then dim the kitchen lights and concentrate on the table lamp for meals. It should be remembered here that the eyes are capable of adapting to a wide range of brightnesses but what they do not like is too great a contrast. So don't leave the working part of the kitchen in complete darkness when sitting at the table, but reduce the light so that the room 'recedes'. Rise and fall are much more convenient since the light can be heightened so that it illuminates the whole of the table for reading or other work and lowered to make for a more intimate atmosphere for dinners. This light should be on a dimmer switch to make it as flexible as possible. Much the same rules apply to an eating area in a living room. If you can extinguish, or at least positively dim, the lights in the rest of the room, the dining area can become a small universe of its own, with its own atmosphere.

Where food is concerned colour is one of the most important aspects. In home lighting quite a large proportion of light from fittings is reflected on one or two or more room surfaces before it finally alights on the table top or other feature it is to illuminate. At each contact with a coloured surface the light is changed by a selective reflection and absorption of colours, so that it finally ends up in a colour at least partly determined by the colours in the room. Sometimes the effect can be quite charming. For

This informal and pleasant eating area is also a conservatory. General light comes in from the wide entrance to the kitchen and is reflected off the pale wall. If you are lucky enough to own a little antique glass beaded lamp like this one, you can use it as a centrepiece for your table setting, adding a couple of candles if you like. The only thing to watch is that no one trips over the lead.

Opposite
A converted mill makes a large
and spacious kitchen/dining
room. The architecture is strong
and the furniture must be bold
and strong as well. The large
pendant glass lamp in white has a
solid, sculptural quality which
complements the rest of the room
and very adequately lights any
activities taking place at the table.

Above
Colour is important and here pink
has been chosen and followed
through very carefully. This is
dining for someone who takes
eating seriously and likes a formal
restaurant atmosphere. The
tablecloth is cotton and matches
the napkins. The china is delicate
and pretty and the lamp prettily
elegant with a pink edge.

instance, if you are using a GLS bulb, with its
yellowish colours, through a pink shade onto
an orange table cloth the result can be warm
and cheering. However, you can also get ex-
tremely unpleasant results, for instance where
the general decor of the room is a deepish green
or blue. The light from that same GLS bulb will
now react with the decor to make meat look a
dirty brown and other foods equally unappetis-
ing. Thus both decor and lighting must be
thought out very carefully in any room where
people are going to eat. This is another good
reason for the rise and fall lamp in dining rooms,
since the light, shining directly downwards, as
it does, will not have to be reflected off walls
before it reaches the table.

If you are not sure what the results will be
from a particular colour, stick to pale walls and
add your deeper colours in small concentrated
areas by judicious choice of particular objects;

A masculine and very spare room whose severity is softened by the large mirror with the ornate carved frame in which are reflected two wall mounted uplighters and the asymmetrical pendant over the table. Pale walls and the pale polished wood of the table help to brighten what might otherwise be a somewhat dark space.

curtains, cushions, flower arrangements; or try coloured bulbs – at least you can change those if the result turns out to be intolerable. These colour rules apply especially to dining rooms per se which don't have to share their uses. Here you can use the dual lighting system not to conceal the rest of the room, but to highlight certain aspects of it. Any good wall or ceiling moulding should be made the most of, of course.

But the dining room is often an excellent place to display paintings or other works of art (beware of the heat, though, if you use a conventional spot lamp; you might do well to install a low voltage light here). You will also probably have a sideboard or table with a bowl of fruit, waiting cheese board and so on which you could highlight to great effect with a small spotlight or some other downlighter.

Work areas

There are many work areas in a home, including kitchens and children's rooms (since toys are work to a young child), but their needs are covered elsewhere. In this section we are talking about desk areas; for writing, reading, word processing, bill paying and also such things as sewing, or indeed anything which has the same sort of requirements involving close eye work for someone seated at a table or desk. A desk may take very different priorities in different homes. It may be a corner of the kitchen table to write out bills, or an old-fashioned bureau with letter writing equipment; it could be a teenage study area, or a complete drawing or writing office for freelance writers to earn a living at. Different though the emphasis may be, the lighting requirements remain the same.

Professionals talk about lux, lumens and footcandles, which are ways of measuring the amount of light (and information on these can be found on page 140), but in practice what you need for close work is at least 200 watts of incandescent or 400 watts of fluorescent light directed onto the work area to give the eyes the most comfortable and efficient light, with a good general back up light elsewhere. It is important, first of all, to put the desk in a convenient position. It should be in a part of your home where it can be undisturbed by the front door bell, children or friendly neighbours. In nearly all instances the desk should be placed against a wall and certainly not in front of a window where the view may be spectacular, but it will be hard to concentrate. Not only that, but light

One low voltage standard uplighter is all that is in use in this basement work room. The white walls and ceiling, the intensity and the whiteness of the light, all combine to provide adequate working light, though the room might be cheered a little by some extra light at the far end of the table.

Opposite
A small, very narrow room has to
hold the filing, paperwork, word
processor and printer of a
professional writer. At the centre,
the ceiling is tall enough to hold a
pendant in a paper shade which
does not reflect into the screen.
On the desk the angled lamp can
shine up when typing, down
when writing.

Right
This charming desk top is
certainly for occasional studying,
or writing letters, rather than for
serious work. The whole space is
lit from a ceiling light and the
cardboard cut-out Chrysler
building with all its lights
twinkling is just for fun. The
small flexi aeroplane light can be
used for reading, or model
making, but uses the smallest
candelabra bulb so is only good
for close work.

from the window alters all the time, being dull
some of the time and brightly sunny at others,
so that your interior lighting will have to be
constantly changed to adjust to it.

The walls should be matt and covered with
emulsion paint, wallpaper or other textured
covering and preferably plain, not patterned, as
the workroom already has enough going on
with shelves, pinboards, papers and equipment
without any extra distractions. Desk surfaces
should also have a matt rather than a shiny or
highly polished surface, from the lighting point
of view. As in any other part of the house, and
with all activities, you will need sufficient
general lighting to see the room and its contents. If you don't have this the contrast between
the brightness of the work area and the darkness of the other will be a strain on the eyes. The
best light of all to work by is good daylight.
Having said that, most people are not, in fact,
blessed with windows which receive good light
all day. Either buildings or trees will block the
light, or the sun will blaze in and have to be
screened off. And, of course, all rooms will need
artificial light in the evenings and at night.

Fluorescent light is still used a good deal for
office work because of its white shadowless
quality. But in most situations even the warm
'de luxe' colours are inclined to give a soul-less
feeling to a room, unless the light is clearly
concealed behind baffles and fixed just to shine
down onto the work surface with separate
tungsten fittings providing a warmer, more
friendly back-up light elsewhere. One advantage of fluorescent lighting is that the tubes last
longer than conventional bulbs and they use
less electricity for the same amount of light. An
alternative to fluorescent lighting is ceiling
track lighting. This must be carefully placed if
the spots themselves are not to shine directly
into the eyes when you are not at work. The
height of the ceiling, the positioning of the track
and the angling of the fittings is all-important,
as well as the choice of wide or narrow angle
bulbs. You will need a fairly wide beam to light
up the whole work area and a narrower one if
you are shining it onto a small pinboard. One
trouble with spot lamps is that they generate a
great deal of heat as well as light. Another is
that you can't reach out and adjust the direction of the lamp as you work, so spot lamps
cannot be substitutes for desk lights.

For general lighting, a small office is one of
the few places where the ubiquitous Chinese

paper globe really comes into its own, providing glare free ceiling lighting and adding a touch of the friendly and informal to the office atmosphere. Having got the background lighting to your liking, you can concentrate on the all important desk lamp. Now to the lamps themselves. Desk lamps should stand so that the lower edge of the shade is about eye level with the person sitting at the desk. A 75 watt or 100 watt incandescent bulb with reflective interior concentrates the light and gives the impression of a larger bulb. But there are many desk lamps to choose from, and in the end the person whose desk it is will have a better idea of what he or she wants than anybody else. The most popular lamps among architects, journalists, designers and others who work at a desk or drawing board is the angled desk lamp such as the Anglepoise or the 1001. These lamps are angled with springs which keep them in place once positioned. The heads are very flexible and they can be used as downlighters, shining down on the printed or written page, or as uplighters, illuminating the work area by reflected light from the ceiling (which is good for computer work). There are table mounted versions of these lamps as well as clamped versions, and a floor stand is also available. The head and arm section simply slots into whatever base you choose and its flexibility makes it universally popular.

For people who use a desk more for occasional studying or writing letters, say in a library or study area, 'student' type lamps like the old oil ones are picturesque and provide good working light though they cannot be angled. They have an old-fashioned look which blends well with reproduction furniture and a study or library atmosphere.

Lighting for small screens

Many families nowadays have at least one computer which involves using a small screen, and many have a serious word processor or micro computer for work. These need rather careful lighting as the images on the screen are already fairly tiring on the eyes and it is important not to make them worse by adding reflections and glare. Most computer screens have a poor light output. Very bright lighting in the room will simply overpower the images on the screen. However, since light levels from natural sources (the window) vary from hour to hour, you want to make sure you can adjust the light to make it as comfortable as possible for you while you work. Dimmers fitted to your room lights will help considerably. Glare and reflections on the screen often exist without the operator even realising they are there. The brain learns not to take them in. Nevertheless, this is in itself extra work for the brain and can be distracting. Check on your own screen – can you see the ceiling fixtures reflected in a corner? Is there a spot of glare anywhere? They are fairly easy to get rid of if you can dim your lights and also adjust the beam, or place the computer or the light where they will not get in each other's way at the outset.

Fixed spots and ceiling lights are the least flexible, though spots on track can be adjusted – and if the ceiling is high enough and the room small enough, any ceiling fitting will probably be out of sight from the screen anyway. Louvred fluorescent lighting is suitable for such situations. Many lighting manufacturers offer such fittings for home use, giving high light levels for low power. Shiny objects such as mirrors and glass covered paintings should be eliminated from the room. You can tilt a painting or mirror slightly to cast the reflection upwards, but this always looks a bit odd. If you can't afford to have a recessed ceiling light, an overhead pendant lamp with a globe shaped lampshade will give adequate general lighting without casting reflections on the screen or shining so brightly that it eclipses the images thereon. Spot lights are not ideal but if the worst comes to the worst they can be concealed with diffusers or baffles.

Some experts tell you that watching a small screen for hours will harm your eyes; others say it won't, that the eyes can work efficiently for long hours under all sorts of lighting conditions and using a screen cannot damage them. The truth probably lies somewhere in between and whatever the experts say, many people do find that their eyes get tired after long exposure to a small screen, as they get tired after reading small print for hours. The eyes may become red and itchy, and there may be difficulty in concentrating. One thing you can do is work only for one or two hours at a stretch and then have a rest. Do some exercise, go for a walk. Roll the eyes slowly round in your head a few times, first one way then the other, before returning to the screen. Three hours of close work is quite enough at one time. However, this tiredness may be as much the fault of the position you are sitting in as of the actual screen. Lack of concentration may be caused by general tiredness. Nevertheless, the better the light for working, the better it must be for your eyes and it certainly makes for very much more pleasurable and comfortable work.

Bedrooms

A bedroom is a much more personal room than a living room and on the whole, easier to light satisfactorily because there are fewer people using it and fewer activities taking place there, though it is surprising how many things you do have to light, when you think about it: making up, wardrobe lighting, bedside lighting and so on. Many people like their bedroom to be unashamedly pretty, very feminine and fresh with a great many floral prints and plenty of flounces and frills. There is no doubt that such a room is far removed from humdrum, everyday work or family life and can be soothing and refreshing. The other side of that coin is a room which is dark and mysterious, slinky and sexy, with silk sheets and satin drapes and a wildly wicked atmosphere. Both offer an escape into fantasy.

However, people living in small homes may find that to type cast a whole room in this way can be a bit wasteful of space and limit possible other uses for the room. With imagination (and discipline about hanging and care of clothes) a bedroom can be so much more than just a room for going to bed in. It can be a private living room, somewhere to get away from rumbustuous family life and to sit quietly in a comfortable armchair and read a book. But even more than an extra sitting room, a fairly big bedroom can be used as a workroom – why waste its daytime potential? A small desk or work table as an extension of the dressing table can be used for diary or letter writing or, if the room is large enough, as a sewing area.

Then there is the bedsitting room, in which the living and the sleeping quarters have an

A pretty touch next to the window of this Georgian bedroom is the vase shaped ceramic base with its simple shade next to a heavy patterned curtain, caught up in a bow. Odd touches of lighting like this are what can give a room its charm.

equal importance because there is nowhere else. The bedsit is a time honoured idea and the Japanese custom of rolling up the bed or folding it to turn it into seating during the day, has caught on in a big way for people in small houses or apartments. It is also an excellent idea for teenagers who may want to spend most of their time away from their parent's eagle eyes, old-fashioned tastes and boring music. How easy it is to use the room as a living room as well as a bedroom, will depend not just on its size and layout but on how you decorate and light it.

Rooms with four poster or canopied beds are unquestionably bedrooms, since the bed can't help but dominate. The only place I know where this is not true is in a Parisian apartment in which there is a tiny, uncomfortable metal-framed folding bed, with a white linen canopy, said to have been used by Napoleon during his campaigns. This is so small and pretty it is kept in the living room and is really more of a curiosity than a true canopied bed. A room filled with one bed and an enormously elaborate

headboard housing television, high fi and tea-making equipment is condemned to be nothing but a bedroom. In fact, in all rooms with a very large bed, everything centres on the bed. Even if it is a divan and used as seating with cushions thrown over it during the day, it will always be the focal point of the room simply because of its size.

Bedside lamps
Whatever you decide to do with the bedroom, the very first rule in any room where people are going to sleep is to make sure there is a light that can be turned on from the bed and one by which people can read in comfort. (Even if you have a convertible settee in the living room, you should make sure there is a lamp in that room which can be reached from the settee when it is acting as a bed.) This means that the beam should shine onto the page without shining into the eyes, either of the person reading or the person on the other side of the bed, who may want to sleep or listen to music on headphones rather than read

Two ceramic bedside lamps take pride of place in this large uncluttered bedroom with its very fresh, pretty greys and salmons. Both have pale shades so that general light is diffused through them as well as being directed downwards for reading.

or even watch TV. One answer is to have two small spot lamps fixed to the centre of the headboard, with each lamp angled slightly outward so that the beam goes only to the person who wants it.

In Christopher Wray's roomy bedroom in south London he has built an alcove round the top end of the bed which houses shelves for bedside books as well as wall mounted lights with Victorian style glass shades, eliminating the need for a headboard. The bedroom is covered in a wallpaper with a large print design in Art Nouveau style.

If the bed is a four poster or has a canopy, then you will need to have lights inside the canopy or drapes, either on a bedside table, wired in to one of the uprights or clamped conveniently near the pillow. For those who find that the peaceful and 'pretty' look is the most relaxing and pleasing, there are lights galore to choose from, such as Victorian glass shades based on the style of old paraffin lamps, which can be attached to wall fittings or to table

Above
Christopher Wray's bed is boxed in at the head so that it can accommodate lights and bookshelves and give a feeling of security. He has chosen frilly glass shades for his all-important bedside lights and they shine downwards to show up the different patterns on the patchwork quilt.

Left
For the pessimistic, a gold bedroom will give a feeling of summer and of sun. This room is very simply decorated with matching curtains, bedcover and wallpaper. It is lit by two small wall brackets with glass shades. The bed is so large one light is hardly likely to disturb the partner on the other side of the bed.

This unusual and very pretty small lamp comes from Japan and is made of cane, bamboo and paper. It gives a charming, unobtrusive light and can hold its own in a variety of different environments. Here, it acts as a bedside lamp in a modern bedroom.

lamps. Frilly fabric shades go with frilly pillow cases, flounces, valances and drapes, and look good on a variety of stands – as long as they are not too severe or too metallic. Heavily draped, curtained and beflowered rooms with heavy Victorian furniture are eminently suited to the rather ornate beaded shades, or Tiffany style lamps and shades, which, although ostensibly from the time of Art Nouveau, to modern eyes merge with Victoriana very well. However, should you want a simpler, bolder, less frilly and more masculine or living room like effect, there are lamps to suit every taste. Ceramic vases or bowls make simple and effective bedside lamps; small flexible table lamps are available in all lighting and electric stores. Modern 'designer' lamps (of which the Italians are still the kings) are available in many variations including those which use tiny low voltage bulbs.

General lighting

General lighting for bedrooms can be gentler and at a lower level than other rooms in the home and the colour of the shades more romantic, using reds and pinks. Ceiling pendants are still popular but they don't really give a very comfortable light for a bedroom, any more than any other room. Strip lighting fitted round the room at ceiling height, concealed by a baffle, will give a pleasant light, washing the walls and making the most of interesting wallpaper. However, there are plenty of pretty wall fittings to create pools of light which are much more restful than top lighting. You can use the dressing or make-up table to create little 'still lifes' with photographs in silver frames or other collections, lit by a ceramic table lamp. Nostalgic and soothing.

Attic bedrooms are exciting to light because you can make something of the roof shapes by uplighters, bouncing light off the ceiling. Table lamps on low tables can also bounce light off a ceiling where it slopes down towards the floor.

Wardrobe lighting

From general lighting we can turn to one of the most difficult lighting problems, lighting of the wardrobe. Most cupboards are not well lit. Either the whole room is lit up like *Close Encounters of the Third Kind*, so that it is not peaceful nor restful at all, in order to illuminate the contents of the wardrobe; or the room is pleasantly lit, but its owner has to feel about in the wardrobe in an effort to find out what's in there. It is possible to put lights actually inside the cupboard, but it is difficult to place them so

Another Japanese style lamp, this time on the floor. Its rather austere black and white colouring and cube shape seems to be much at home with the Japanese shelving and various hi-fi, music and video systems this bedroom enjoys. It gives a fairly efficient diffused light through the white paper shade.

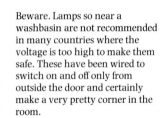

Beware. Lamps so near a washbasin are not recommended in many countries where the voltage is too high to make them safe. These have been wired to switch on and off only from outside the door and certainly make a very pretty corner in the room.

that they don't cast shadows and make things even harder to see than before. You can install strip lighting all the way down the inside of the cupboard on an automatic switch which switches the light on when the cupboard door is opened, and off when it is closed. But spot lamps are probably the best form of cupboard lighting. Set them high enough in the ceiling and near enough to the wardrobe, so that the beam can be directed onto the interior without you getting in the way of it. If you put the lamp on a track, you can add another one or two, to direct elsewhere in the room, or, if you have a fitted range of wardrobes and other storage, direct all the spots to give satisfactory lighting to all the storage.

Mirror lighting

While on the subject of clothes and dressing, in my view every bedroom should have a full length mirror but these are difficult to light correctly. The secret is to light the person and not the mirror. Any light source in front of the glass will simply cause reflections so that you can't see anything for the glare. But if you have a spot light shining from behind the mirror down onto the place where you stand to look into it, you will get an excellent well lit view of yourself. During the daytime, the light will be coming from the window, so place the mirror where the daylight won't shine directly onto the glass, or again, the reflections will prevent you seeing what you need to see.

Study or work area

There are many table lamps to choose from which are less formal than the sort of lighting you might want in a room set aside specifically for working. The original paraffin 'student' lamp is suitable for certain types of work, particularly where reading is involved, but for sewing, painting or other fine work, an angled lamp would probably be more convenient. Again, you might not want the ubiquitous office lamp with its high tech style, but there are plenty of smaller, more elegant angled lamps both modern and traditional, which would be ideal, in particular lamps which clamp onto shelves or uprights and don't take up precious space on the table.

Bedroom switches and sockets

It is important in any room to have plenty of socket outlets so that you can avoid the dreaded Christmas tree effect of plugs and flexes sprouting from a triple plug, which makes for

Very discreet lighting for those who want a four star hotel atmosphere in the bedroom. These downlights give excellent general light, good for mirror lighting and, if on a dimmer switch, flexible for mood changes. Ideally they should be dual-switched from the door or the bed.

overheating and can be quite dangerous. Architects often underestimate the number of sockets which will be needed in a bedroom. But the bedroom in a modern home is likely to have a hairdryer, heated hair curlers, a shaver, radio, TV and possibly an iron which will all need sockets. On top of that you will want to vacuum regularly, you may want to use a sewing or knitting machine, and that is quite beside the number of table and floor lamps you might like to have in the room. Double socket outlets are not much more expensive than single ones and certainly worth having, since you will almost certainly use the extra socket. It is always best to install more than you think you will need. You can, of course, always use extra sockets on an extension lead, but it is both safer and more satisfactory to install sockets in the first place.

There are elaborate headboards which include sockets, switches, phone and overhead strip lighting positioned for bedtime convenience. Such headboards are essentially for modern bedrooms. If you prefer the 'cottage' look, keep your sockets in the wall. Sockets for heavy equipment should be 45 cm (18 ins) from the floor. For hair dryers and shavers which you are more inclined to wave about in the air, 10 cm (4 ins) above table top height is more satisfactory.

As with lighting in other rooms, it is a good idea to have each type of lighting on a separate switch, so that your cupboard, mirror or dressing lights can be switched together, the general lighting together, the work lighting on its own and the bedside lights switched from both bed and door. Do have all these on dimmer switches, which not only helps to change the mood of the room but can save on electricity as well.

The combination of draperies, frills and the naked flames from candles used to cause innumerable fires in homes which were largely made of wood. Nowadays, cigarette smoking is the main culprit in the bedroom – when people fall asleep with cigarettes still alight in their mouths or smouldering on the bedspread. Now that lighting is so much safer, it is possible to re-create the candle-lit effect or to have a high level of light without fear of the consequences. The candle-lit look can be achieved by using simple old-fashioned lamps with candle bulbs, or even just by judicious use of dimmer switches on warmly coloured lamps.

Bathrooms

First let's look at the safety factor. Electricity and water are a dangerous combination and it doesn't take much moisture to conduct electricity, so bathrooms are full of potential for giving lethal electric shocks. In some countries there are stringent regulations about electricity in bathrooms. For instance in Britain, where the voltage is higher than in other countries, portable lamps are not allowed to be used in the bathroom at all and the law says that bathroom fittings should be sealed with glass or plastic covers, leaving no exposed metal that could be affected by steam and/or condensation. So most uplighters, and wall sconces in general, are not suitable for British bathrooms, nor is tracklighting with spots, nor some downlighters. It is also compulsory to install the switch outside the bathroom altogether or to fit a pull cord switch. In other countries, regulations are less strict but it still pays to take care. Basic common sense

Here is a bathroom reminiscent of the Victorian heyday when mahogany ruled supreme. The central chandelier with its white globes gives excellent general light, backed up by the wall brackets and a more ornate pendant in the background. All help to illuminate the many pictures and prints on the walls.

says you should always switch off the power at the mains before changing a bulb, specially with pull cord or two-way switches where you have no means of knowing whether the light is switched on or not. No piece of electrical equipment should ever be touched with damp hands. Your hair dryer, power operated radio or any other piece of electrical equipment are absolutely taboo in the bathroom, at least where they can be reached from the bath or shower. If in doubt, get the advice of an experienced electrician. There are many purpose designed fittings both wall mounted and ceiling mounted, from glass globes to opaque discs and strip lights, which are suitable for bathroom use.

As usual, the general lighting should be tackled first and this is often inadequate in bathrooms. The full effect of carefully chosen tiles and matching bathroom suites is often completely spoiled by dismal and inadequate lighting. Since the bathroom is absolutely full of shiny surfaces, it can easily cause uncomfortable reflections and glare, so bear that in mind as you do your planning. If using reflector bulbs you will find that gold ones produce a warmer feeling than silver, though they will be more difficult to locate in shops. If the bathroom is very small indeed you can combine both heat and light with one infra-red bulb attached to the ceiling fitting. These are not particularly decorative but give adequate general light and enough heat to take the chill off, to make bathing or showering more comfortable in those between times before the central heating is turned on for winter or after it's been turned off in a cold spring. They are also very good for changing a baby under in an otherwise unheated bathroom. For larger bathrooms there are variations of light-with-radiant-ring heater in which you can turn on just the light, just the heater or both together. These are workmanlike and efficient, though cannot be thought of as decorative by any stretch of the imagination. But again, they provide adequate background heat and light and from then on you can build your decorative lighting. Other forms of background lighting suitable for bathrooms are ceiling recessed lights, several low voltage bulbs giving a pleasant and more subtle light than pendants, and globe lights specially designed for bathrooms which can be attached to the ceiling or wall.

Some people do their grooming in the bedroom, others prefer the bathroom. For grooming you need a good mirror, so mirror lighting is all important, but it is hard to get right. If you want realistic, albeit somewhat harsh, truthful lighting to shine on your face while making up, then bulbs round the mirror, as used in stage dressing rooms are by far the most effective, but they must be correctly placed to give light that is both free of shadow and free of glare. There should be no shadows on the face, none under the eyes, nose or chin, either for shaving or making up. The most effective light in these circumstances is an equal level of light coming from the sides and top of the mirror, not just from either side or just from the top. Horizontal and vertical architectural or fluorescent strip lights at either side and above the mirror will work well or you can get a fixture for several bulbs and fit them round the sides and top. This lighting next to and above the mirror throws light directly onto the face and gives a clear unflawed image. Bulbs set further away or further above shine onto the glass creating reflections and dazzle, which are uncomfortable and unhelpful, whereas bulbs shining up from below cast quite deep and disfiguring shadows. Bathroom mirrors are available nowadays with built in electric sockets for two or three bulbs at the top. If you are using fluorescent strips, get one of the de luxe warm white colours for the most natural (and least unflattering) light. Individual light fixtures should have 60 or 75 watt bulbs to give enough light, but if you are using more bulbs together, 40 watt each or even less will be enough. If you don't like the idea of such ruthless lighting there are many decorative wall fittings which, fixed on either side of the mirror will be pretty in their own right and give perfectly adequate make-up and shaving light.

Use mirrors creatively and positively. If they are fixed in certain positions they will double the amount of light and can make a rather poky bathroom seem quite luxurious by doubling its apparent size. Mirrors should, if possible, face the window so as to reflect any natural light coming into the room. A whole mirrored wall next to the bath will reflect all lights and make the room seem quite spacious. Use plastic mirror if you don't want it to mist up. Mirror tiles can be effective, giving a broken up image which may be less demoralising than a perfect one for most human beings.

If the bathroom is big enough you can introduce plants, an armchair, paintings and make it more like a pleasant room than the purely practical one which the space in most bathrooms demands. The survival of the plants will depend to a certain extent on how much light comes into the room and whether it is

Opposite
Oh, the luxuriousness of this wonderful large bathroom mirror with its galaxy of lights! Notice that the bulbs are fixed only round the sides and top. Any below would cast ghoulish shadows over the face and prevent clear vision for making up or shaving. It's worth getting good quality mirror if you intend to light it in this way.

From the bathroom, this pretty, thirties mirror reflects the washing area in the neighbouring bedroom. The broken image adds a mysteriousness to the little indoor garden scene with its sprouting Art Nouveau type lamps.

centrally heated. But for plants like ferns, which like a humid atmosphere, the bathroom can be the perfect environment.

When planning the bathroom lighting, consider using dimmers. With these you can tone down the light when you don't need the full blast on your face. Larger bathrooms are better with two or three sources of light rather than just one. Once you have seen the difference this can make, you will wonder how you ever tolerated a one-light bathroom. Larger bathrooms offer all sorts of refinements to the lighting plan. For a start you can recess the basins, building round them to create cupboards, things most bathrooms are very short on anyway. Under the cupboards and over the basins you can install strip lighting (in some countries this is not legal), very much as in a kitchen above the worktop. This is convenient, efficient and looks inviting.

Opposite
There's no need for a bathroom to look ultra hygienic, as can be seen from this one with its pretty poppy field wallpaper and collection of blue and white plates. What better to light it than an old fashioned gas lamp wall bracket set high in the ceiling, out of the way of the water and illuminating both plates and bath.

Children's rooms

Opposite
Rise and fall pink pendant lamp in a child's room is matched by pink curtains and pink walls. It is practical illumination when lowered for the floor play which goes on in this room. For more general activities like dressing and going to bed, the light can be raised and will illuminate the larger area of the room.

Below
A bit of fun for a young child; a dolls' house street lamp bought in a Milan junk shop, lit by a torch battery, standing outside an old biscuit tin 'Bicky House' designed by Mabel Lucie Atwell, with a Christmas tree candle planted nearby. This may not be serious lighting, but it's enough to keep the goblins at bay.

Most parents understand very well what children need in their own rooms as far as furnishings are concerned. Babies and young children are given warm floors, efficient storage, plenty of shelves and a sturdy work table at the right height for activities or homework. However, the lighting for children's rooms is often ignored and misunderstood, although children need good lighting quite as much as anyone else, even from a young age.

Children's toys are their tasks, through which they learn manual skills and an understanding of the adult world. If they are going to make the most of, and learn the most from, their posting boxes, jigsaw puzzles, matching-the-picture games and drawing and painting activities, they must be able to see. For that they need a high level of general lighting plus any extra lamps necessary – over the writing space, for instance – to act as task lighting.

General lighting for young children should be bright and revealing. They and their friends are likely to make full use of the whole room and of all the floor space when they are playing. They (and their parents) need to be able to see clearly into every corner as well as in the middle of the room. Such lighting could be provided by a single ceiling pendant with a high wattage bulb, up to 150 perhaps, and a pretty but not too obscuring shade, or one or two tracks with spots set into them, depending on the size of the room. For older children it is important that this general lighting should be boosted by various added lights for the different activities a child undertakes. Children often sleep, work, play and entertain friends in the one room, and individual lights for doing all these things should be chosen with care. Take the bed, for instance; very young children may enjoy a pendant lamp with a pearl globe bulb, a crown

silvered bulb or a standard GLS bulb shaded by a fabric pendant shade. There are many fun shades on the market such as gathered ones with teddies or some other sort of furry friend dangling from the fitting (which a baby could look at from the cot without hurting his eyes). A light over the bed is useful if the child wants to sit up and play with toys or read. The light can be controlled by the parent from the door, so there's no chance of him harming himself by poking and prying, or sitting up and reading after lights out.

Older children should always have a bedside light which they can switch off from the bed so that they can read or play comfortably before settling down to sleep. For boisterous children such lamps may be best fixed to the wall so they can't be knocked over. Bunk beds should have independent lights of course, so that one child is not prevented from reading in bed while the other sleeps. And, by the way, although bunks may look neater and more aesthetically pleasing if close together, the space between them must be tall enough to allow the child in the bottom bunk to sit up and read comfortably. Some bunks are rather meanly proportioned in that respect so that the child below cannot sit up straight.

All desk areas in the house should be provided with an adjustable lamp. Children's homework areas are no exception. Children have plenty of homework from the time they go to school, but long before that they like to paint and draw, cut out and stick, and they need a good light for those tasks. Apart from obvious things like model making and sewing there are all sorts of other activities – anything from stamp collecting to making collages which need good adjustable lighting which won't cast shadows, if the child is going to be able to work enjoyably and

Part of Christopher Wray's own collection of Victorian ceramic night lights for children. Owls feature predominantly, presumably because they are night birds, but anything from elephants to dogs and squirrels were used as well. These were paraffin lamps and the creature held the paraffin.

Opposite
The modern version of the night time animal friend. Animals such as this plastic rabbit have an electric light fitting inside so that the whole creature glows with a friendly warm light. Geese, ducks, sheep, dogs, pigs and other animals are all available. This one looks as though it's about to hop into the Beatrix Potter landscape painted on the wall.

efficiently. You can provide a 'serious' angled lamp such as the one described in Work Rooms or a smaller table lamp with a flexible stem. Some brightly coloured fun lamps can be entertaining and still quite practical for the home-working or model making child. If the table is small and you don't want it cluttered with unnecessary objects, use a clamp or a clip lamp fixed to the edge of the desk, work top or to the shelf above. The track and spots are perfect for the nine-plus age group, to light up posters and pinboards etc.

One of the important considerations for children's rooms is the lighting that keeps them company at night. Most children need the assurance of a little night light to help them get to sleep without fear and in case they wake up in the middle of the night. This doesn't have to be a bright illumination, indeed too bright a light would be disturbing. There are a number of ways to provide such gentle night lighting. Wax nightlights were the old fashioned way of giving children assurance but they didn't last the whole night through and were superceded by tiny glass lamps in the shape of animals or birds – owls which glowed in the dark were popular. Now there is available a whole menagerie of life-size geese, rabbits, ducks, pigs and sheep to keep a child company at night. Children really love these lit up creatures and often make friends of them and chat to them at night. If you are going to place them on the floor, be sure you choose safe ones, where the bulb socket is inserted in such a way that the child can't pull it out. Other individual lamps offering a friendly glow in the dark include ceramic pixie houses and mouse homes, old women living in shoes and suchlike which light up inside, giving a cosy feeling of companionship and warmth. Then there are friendly men in the moon orbs which glow like the real thing. In fact there are a great many imaginative, dim but reassuring lights designed to make night time less anxious for the timid child. And, of course, there are dimmer switches to control any or all of the main lights in the room to give a gentle overall level of light which won't disturb a child's sleep but will allay fears. Gradually, as children grow older and sleep longer, such lighting will probably become unnecessary, though during periods of illness it may be a good idea to have some form of gentle night lighting available.

In your children's rooms it is most important to remember safety rules. There should be absolutely NO trailing flexes, nothing easily knocked or broken. When installing sockets, set them in at table height so that young crawlers and toddlers won't be tempted to poke their fingers or knitting needles into them. Where floor level sockets already exist you can buy plastic, clip-on socket covers very cheaply, which is a sensible precaution for sockets which are not in use all the time, at least until the child is old enough to know better than to play with them.

Unfortunately, children are prime targets for badly made goods, since if something is pretty enough or advertised enough a child will want it, and it can be very hard to say no. However, where lighting is concerned, badly made fittings must be absolutely taboo. Metal lamps may be badly insulated and become live. Plastic lamps may break or simply fall to pieces. Always look for safety symbols when buying lamps, or buy from reputable manufacturers and stockists. Old fittings are also a potential danger.

Out of doors and leisure

Left
This is an original front door lamp designed for paraffin. You can still see the metal reflector at the back which helped create more light. The glass had to be very well sealed to protect the flame from the outside elements. Such lamps are now available for use with electricity.

Right
A small fairy mushroom growing among the plant pots gives a very adequate light in the most charming way, without being too sentimental or 'gnome like'. This is an Italian fitting giving a clear, white light which looks good in most gardens.

A little low wattage goes a long way. The first priority when considering outdoor lighting is the front door. Lighting the outside in general is of prime importance simply for safety's sake. The better lit the outside, the less likely you are to be burgled, since it is much more difficult and daunting to creep unbeknown to a well lit house than one with concealing shadows and darkness. It helps, no doubt, to have neighbours, too, but if you have none, there's even more reason to make sure you have adequate lighting. The second reason for good front door lighting is for the convenience of everyone living in the building. If the light is such that you can find the front door key and insert it easily into the lock, then coming home is an altogether pleasanter prospect. Some people like street lamps, others carriage lamps, and others no-nonsense bulkhead lamps or glass globes. The choice will depend on your own tastes, and no doubt, to some extent, the style of the house.

Although most people do take some trouble to light their front door, very few really make use of lighting to make the most of their garden after dark. Yet the shapes made by shrubs, bushes and trees in the garden can be both mysterious and intriguing if well lit, and can lengthen the time the garden can be used, just as much as solar heating increases the use of a swimming pool. It is important, first of all, to choose lamps which are designed for use out of doors and then to design the lighting scheme with subtlety and care. Of course, if you have a summer house, outdoor table and chairs on a patio, a sculpture or an interesting aspect of the main house, you will want to make the most of these with your outdoor lighting. You can use fittings to take a PAR 38 bulb, perhaps a coloured one – but beware of using too much colour, which can make a garden scene gaudy rather than enticing. If your garden is not blessed with any particularly interesting piece

of architecture or furniture, there are plenty of ways of lighting up the most mundane shrub so that its form and leaf shapes show up to best advantage, lifting the whole garden out of the ordinary.

There are various fittings to take a PAR 38, from wall mounted ones to those with a long spike which you simply plunge into the ground. This has the advantage that you can pull it out and move it about to light different plants as they bloom, or simply ring the changes with your garden lighting in rotation. Both spot and flood versions of the PAR 38 exist so you can play around with effects until you get what you want. The PAR gives a yellow light which is particularly effective in autumn, showing up the reds and yellows of the leaves as they turn colour. More powerful lighting is obtainable from sodium and mercury vapour bulbs. Sodium lighting is the very orange lighting often seen over motorways and as street lighting. It can be a bit brash for the small garden, but effective for large stately homes and their grounds. Mercury vapour lamps give a blueish light much used on larches, pines and other evergreen trees. It is quite possible to use all three to give contrasting effects, but take care not to over-do it, or you will lose sight of the garden in the welter of lights. The choice of reflectors will dictate the angle of beam. Position the lamps facing away from people and towards the plant or tree to get a good view of the foliage without being disturbed by the glare. Low voltage fittings using 12 volt PAR 36 bulbs are more subtle and many people prefer them. It is easier to hide them because of their small size, behind plants, pots, smaller shrubs etc. A choice of narrow spot, wide and very wide beam makes them suitable for nearly all situations. You will, of course, need a transformer, as with any low voltage lamps, and one can be used to control several fittings. However, you must find a safe and sheltered position for it and it will need a long cable to reach all the fittings. You will need a professional installer to make sure all is safe.

Paths and patios may need lighting if they lead from the garage to the house and vice versa, or if the garden is a large one and people will be walking in it at night. Luminous bollards may seem rather formal but are quite effective. Tall lamps, again rather formal, are very suitable for town gardens, meticulously landscaped gardens, those made up in the main of pots and for square borders between slabs of paving stones. Uneven paths and steps should be lit particularly carefully, with the light shining down so people can see where they are going.

There is a whole range of outdoor lights which are intended to be seen as sculptures in their own right, and that aspect is as important, or more so, than the practical side. There are tall, mushroom shaped lamps, small stick lamps, carriage lamps and street lamps.

If you are lighting a large area of wall or hedge, use two spots angled from opposite sides so that any shadows cancel each other out. Ornamental pools should be lit from under the water if possible. It's best to use low voltage sources and, as always, they should be installed professionally, as indeed should most outdoor lighting. An alternative to low voltage is weath-

These lights with their sturdy peg legs and inclined necks are excellent for lighting shrubs and undergrowth, and for leading people up the garden path, perhaps from the garage to the house, or other walkways after dark. They are elegant and unobtrusive.

erproof floodlighting, casting light over the whole surface of the water from just above the surface at the side of the pool. It seems a great pity not to light a fountain if you have one, since the combination of water and light on a dark night can be very exciting. Fountains can be lit by narrow beams set at the base of each jet and shining along it so that all the light shines through the movement and catches the droplets as they fall. Coloured lights can be very effective.

Swimming pools are something else again. If they have an attractive tiled hall, then the lighting should be used simply to highlight the colours and scenes depicted. If the hall is undecorated, or for outdoor pools, the lighting is most effective if set underwater, though this is only really practical if installed at the same time as the pool is built. Most pool lighting consists of 50 watt PAR 36 or 300 watt PAR 56 bulbs positioned to shine down along the length of the pool from the deep end – a dimmer on the lighting is effective. Choose a fitting which can easily be removed and cleaned and make sure any light you choose is really designed for the purpose.

For outdoor lighting, keep your main switches indoors to give good and easy control. Make sure everything is weatherproof and waterproof where appropriate. Keep fittings and cables regularly inspected and free of debris such as leaves and stones etc. Always get professional advice for outdoor work and use a professional electrician. Don't let flexes and wires be laid where you will want to use a lawn mower or cultivator.

Apart from outdoor lighting, there are indoor leisure activities to consider too, which are seldom adequately lit and which would give everyone so much more pleasure if they were. If you are lucky enough to have a 'rumpus' room for teenagers, make sure above all that lamps are indestructible and will withstand the sort of rough and tumble which may go on in there. Ceiling recessed spots would be excellent. Adjustable spots on track would be good for a dart board or other wall-based game. A table tennis table should have downlighters of some kind shining on each end. A table which is well lit one side and dimly lit the other makes for a poor game for each side. There should be adequate light over any table used for cards, jigsaw puzzles, chess, draughts and so on. Jigsaws are difficult enough as it is and can be made twice as daunting and far less appealing if you can't see what's on the pieces. If you take your leisure seriously enough to install a billiard table, do put in specially designed billiard lights. This makes all the difference to a good game and to the look of the room.

Opposite
A concealed light source is used here to silhouette a Victorian cast iron circular staircase, surrounded by climbing and rambling plants. It creates a truly fairytale atmosphere in an otherwise very plain brick courtyard.

Special effects

Wonderful, dramatic and really surprising effects can be achieved with lighting. You can use such effects to lead the eye away from what you don't want it to see, or simply to startle, surprise and amuse. For instance a group of neon cacti on the floor by a fireplace can look wonderfully out of place and zany, but will also disguise the fact that the fireplace tiles are ugly or broken. Lamps can also be used to complement an existing feature in the area where they are placed. Luminous globes or opaque glass 'vases' can be used in this way; for example, contributing to the set piece of a

fireplace so that the fire-surround and mantelpiece act as a frame for the light. Many modern lamps have been designed specifically as sculptures. The giant sized light bulb sitting in the middle of the room, for instance, is a splendid object in its own right, a good joke and a useful light source.

Plants are eminently suited to special effects, with their wealth of different forms and shapes. You can add glowing pond lily lights to an already jungly proliferation of green plants, concealing the light source completely by placing it low down, preferably on the floor, and

An almost filigree metal palm tree next to a spiky indoor plant has tiny bulbs in every branch. The resulting shadows spray out onto the wall so that what at first seems like a simple object has a thousand different facets.

Opposite
If you think of fluorescent, as a dreary white, think again. This is just one of many different and very bright colours available. The tubes can be propped up against a wall, laid on the floor, placed on a shelf (or crunched between a tiger's jaws).

If neon is used, it should be used with courage and flamboyance. In this Aztec bedroom, with its brightly coloured screen, bedspread, frieze and rug, the shocking pink flamingo with its shining green grass feels perfectly at home.

Opposite
This is a truly 'over-the-top' effect; a parachute canopy dotted with tiny Christmas tree lights. It even has a kite hidden in its folds. The bed is full of music and there is a fan the size of a headboard. Perhaps this is the modern young person's answer to the four poster bed.

pointing upwards to profile the leaf shapes. Alternatively, position a spot lamp in front of the plant facing away from the viewer, giving a completely different effect again. You can use simple globes of directionless light in conjunction with plants, which give a sort of moon effect, calming and mysterious.

A lamp can be used to bring together an otherwise motley collection of objects. Many of the most modern lamps sold today are special effects in themselves. Painted neon tubes are available in a number of interesting shapes, designed to sit on the floor, which give a charming light, very different from the functional, stark light we associate with them. Fluorescent tubes leaning at an angle like drunken walking sticks give a science fiction

atmosphere and are at their most effective in a large room with stark ultra-modern furnishings. Decorative fluorescents are available in over fifteen colours, some very deep and beautiful. Other uses for fluorescent lighting have been discovered as well, particularly in the circular, pastel coloured tubes which can embellish the ceiling rose of a tallish room.

Very charming effects can also be achieved from table lamps with traditional shades used as downlighters. If you stand such a light on a small table, it is possible to arrange a fascinating little display beneath which will be devastatingly illuminated by the light above. The display can be anything from dried flowers to framed photographs, shells, or other and various carefully chosen objects. You can create a stark

This little group of miniature cacti on their gourmet hamper is given a warmth not usually associated with prickly plants, by the glow coming from the red nylon bag. The bulb used inside is a low voltage SL which does not get hot, so it is quite safe.

elegance by using blues or blacks and whites, or a warm friendly corner by the use of yellows, reds and oranges. Very special effects can occasionally be achieved by using light to show up the colour of a wall. There are certain situations where this can be particularly startling. For instance on the poppy-red wall of a library, where the colour together with the books and the light give a splendid feeling of luxury and studiousness all in one. Special effects can also be produced in the way you light collections of glass or sculpture. Glass should be lit from behind, so that the light shining through catches all the colour and transparency. Glass shelves in front of a window make an excellent display case. You wouldn't

want to do this in a room with a remarkable view, but if the window faces onto a dull courtyard or street, and doesn't contribute enormously to the light in the room, it may be better to utilise the window in this way and get pleasure from it, than waste it completely. Sculptures are more difficult to deal with. The perfect solution is to have a wide enough beam to highlight the whole of the form, but not so wide that it takes in the space beyond it, and not so narrow as to catch only a detail and leave the rest in obscurity. So the beam angle and the distance from the object are both important. On top of that you must make sure the light doesn't shine into the viewer's eyes; so the choosing, placing and angling of the light are all crucial.

THE PRACTICAL SIDE

Light technology has gone ahead in leaps and bounds since
the electric light was first invented and lighting equipment has
improved with it. The plethora of different bulb shapes and
wattages, of dimmers, timers, controls and switches can be
confusing. This guide covers those most easily available and
gives some useful addresses.

A guide to light bulbs

Incandescent – general purpose

GLS (General Lighting Service): this is the standard bulb used for general lighting in the USA and the UK. You will find it in most table and floor lamps, pendant lamps, wall and ceiling lamps and fittings with diffusers and shades. It should have a life of 750 to 1000 hours.
Base: Edison Screw, Bayonet
Watts: 40 to 150 for domestic use.
Finish: clear, pearl, white, daylight and various colours.

Mushroom GLS: the same as GLS but better for positions where the bulb is exposed.

Globe GLS: as above.

GLS strip or tube: tubular incandescent bulb, particularly useful for giving warm light round mirrors.
Base: Edison Screw, Bayonet, Small Edison Screw, Candelabra Screw, S14d, S14s, S15s, disc
Watts: 25 to 120
Finish: clear or pearl

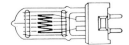

Incandescent – display

Crown silver: a GLS bulb whose front is coated with silver or gold in a clear or pearl bulb. The silvering directs light upwards where it is intensified and the beam is reshaped by the reflector. Use it in display fittings, where it is used to reflect and redirect light.
Base: Edison Screw, Bayonet, Small Bayonet
Watts: 40 to 100

ISL 80/95: internally silvered bulb, used as task lights, downlights, wallwashers and soft accent lights.
Base: Edison Screw, Bayonet
Watts: 40 to 150
Beam angle: Flood 78°, Spot 26°

ISL 125: use as above
Base: as above
Watts: 75 to 300
Beam angle: Flood 76°, Spot 22°

PAR (Parabolic Reflector

Lamps): all PAR bulbs are made of two pieces of glass fused together. One piece is the reflector, the other the front lens. The way the lens is patterned determines the beam angle. This lamp can be used out of doors. PAR bulbs are also available in tungsten halogen versions.

PAR 38: use in downlights, wallwashers, accent lights, outdoor and patio fittings.
Base: Edison Screw
Watts: 75 to 120
Beam angle: Flood 30°, Spot 16°

PAR 46: for outdoor fittings or as accent lights in high ceilings.
Base: Medium side prong
Watts: 200
Beam angle: Medium flood 11° to 26°, narrow spot 9° to 13°

PAR 56: use as above
Base: GX 16d
Watts: 300
Beam angle: Wide flood 19° to 42°, medium flood 11° to 23°, narrow spot 8° to 10°

PAR 64: use as above
Base: Extruded Mogul End Prong
Watts: 500
Beam angle: Wide flood 19° to 58°, medium flood 9° to 24°, narrow spot 7° to 10°

Incandescent – tungsten halogen

These are incandescent bulbs filled with halogen gas which give a more efficient, whiter light and have a life expectancy of 3000 hours.

Linear filament bulbs: give even light distribution. Any over 500 watt should be used in a horizontal position. If being used to replace a GLS bulb, make sure the fitting can cope with the higher temperatures caused by the halogen.
Base: R7s, minical, Edison Screw
Watts: 100 to 2000
Finish: clear or pearl

Incandescent – low voltage

Low voltage bulbs all operate at under twenty-five volts and always need a transformer to lower the voltage from Britain's 240 or America's 110 mains voltage.

Bare lamps: these are tiny bulbs used in more and more modern desk and table lamps, as designers' discover their potential. They are also used in track, recessed and accent lights. They must have a reflector.
Base: G4, GY6.35
Voltage: 6, 12, 24
Watts: 5 to 1000

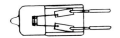

PAR 36: these have a life of 1500 to 2500 hours. There are many bulbs now available, some with dichroic reflectors which divert the heat out at the back of the bulb, thereby making the light beam 60 per cent cooler.
Base: M4 Screw
Watts: 25 to 100
Beam angle: 3° to 60°

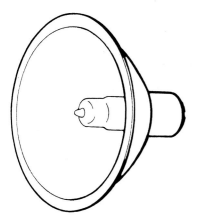

Reflector lamps: there are many different versions available, thus:
Bases: GX5.3, G4, B15d
Voltages: 6, 12, 24
Watts: 20 to 75
Beam angles: 30° to 50°

Irridescent – miscellaneous

There are a great many decorative incandescent bulbs on the market, used for their interest and entertainment value rather than for the light they will produce. Some use colour, some use shape (candle shapes), some can be as small as a snowdrop bud and others as large as a tennis ball. There are flickering bulbs to emulate gas lamps, candle bulbs, bulbs with decorative light-up interiors and so on.

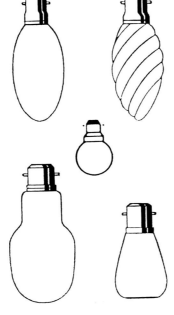

Fluorescent tubes

Straight tubes: these range from 6 ins (4 watt) to 60 ins (90 watt) lengths and $1\frac{1}{2}$ ins diameter (14–40 watt) and $2\frac{1}{8}$ ins diameter (82–90 watt). There are a variety of colours to choose from.
U Shaped tubes:
Watts: 35 and 40
Circular tubes: used for ceilings

Fluorescent – miniature

These are small, compact tubes which can be used in the home, though they are awkwardly shaped for most household purposes. Most have a low level of light.

PL: double tube, suitable for halls. They require control gear in the fitting.
Watts: 5, 7, 9, 11

SL: this has been manufactured with its own small transformer. Lasts up to ten times as long as a GLS bulb.
Base: Edison Screw, Bayonet
Watts: 18

2D:
Base: with a special adaptor it can be plugged into an ordinary bayonet cap socket.
Watts: 16, 28

Controls, dimmers and switches

A dimmer gives great flexibility; enables you to accentuate different areas of a room at different times of the day, or for different purposes. Dimmers are easy to install and can be used on table and wall lights, as well as ceiling fittings. By dimming a lamp you not only lower the lighting level, but thereby save energy and may increase the life of the bulb.

The easiest lamps to dim are incandescent (tungsten) and tungsten halogen, for which you can buy dimmers which operate as light switches at the entrance to a room. Special low voltage dimmers are also obtainable for low voltage lights. If you are having your complete home designed by a specialist, and this is an excellent thing to do if you can afford it, you can have a remote control dimmer, so that you can switch on, off or dim any light in any part of the house – or use a hand held dimmer, say in the hall or in your bedroom.

Other useful controls are time delay switches for stairs or long passages. Though these are useful in converted houses and apartment blocks, they can be irritating in one's own home, unless there are teenagers who simply refuse to remember to turn out the lights (a dimmer, however, can be used so that you can keep the light on at a very low level during the night, which is safer and more energy saving). Mechanical timers can be plugged into a wall socket set to switch the light on and off at preordained times. These are useful for waking up in the morning and can be set to turn on and off several times a day. Photo-electric cells are available which are waterproof and can be preset to turn on porch lights at dusk and turn them off again at dawn. And of course, there's no reason why you should not have several lights operating off one circuit, which means you don't have to go round the house switching off dozens of switches before going to bed.

Care and maintenance

Cleaning

Planning and installing your lighting, though important, is only the beginning of making sure you have good illumination. If you want to get the best results you must maintain and clean your lamps and fittings regularly, say once a month, since the light attracts moths and flies and the heat attracts dust. Frequent cleaning means they won't get so dirty that it becomes impossible to get them really clean. This is specially true of fabric lampshades which shouldn't be washed and are difficult to dry clean. Regular cleaning will also ensure that they operate efficiently and that the light is not prevented from getting out by dust and dirt.

Obviously it is worth choosing fittings which will be easy to clean, especially those which are going into the less accessible areas of the home, such as on staircases, landings or anywhere with a high ceiling. Electricity is, of course, a lot cleaner than paraffin and oil lamps, whose smoke always leaves dirty deposits on walls and ceilings. Users of oil lamps used to protect their ceilings by fixing glass smoke catchers above the lamps. People who like to use such lamps for dining, for instance, can still buy them quite easily. If you do have such a lamp, maintenance is important, not just cleaning the exterior of the lamp but also keeping the wick trimmed. Although electricity is cleaner, fittings still become dirty surprisingly quickly, especially if there are heavy smokers in the house. Tobacco smoke covers shades, fittings and reflectors in the same way as it used to coat the fingers and thumbs of smokers, before the days of filter tips.

Pendant fittings need regular attention, bowl lights can soon become filled with dead live-stock. Always switch off the light first, at the mains preferably. It is quickest and easiest in the long run to remove the fitting. But a word of warning here: some fittings, once put together are not easy to remove and you may be well advised simply to rub over the whole thing with a damp cloth.

Fabric shades are not usually washable. Even washable fabrics may not have waterproof linings, and if you do wash them, the metal frame may become rusty, so on the whole washing of fabric shades is not advisable. They should be dusted regularly, and can be rubbed over with a 'bread ball' or covered with fuler's earth and brushed off. Silk lampshades should be cleaned professionally.

Parchment should be wiped with a cloth wrung out in vinegar and water. Leave it to dry before touching because it becomes very delicate when wet.

Acrylic should be wiped with warm soapy water. Use a very small amount of anti-static polish to prevent dust being attracted back immediately. If the acrylic is scratched, use metal polish to conceal the scratches.

It goes without saying, I hope, that you should allow fittings to dry before replacing them. Leave them on a warm stove or play a cool hairdryer on them. Aluminium reflectors should be wiped with a clean, damp cloth or sponge. Make sure all detergent is rinsed off and buff up afterwards with a soft cloth, or use metal polish.

Glass shades should be washed at least twice a year. They can be cleaned with a commercially available product or a very small amount of gentle washing-up liquid, such as Fairy Liquid.

Chimneys of real oil lamps should be washed in mild detergent and warm water, and cleaned with a chimney brush (like a little bottle brush). In the old days every large house would have its 'lamp room' in which one member of staff would spend all day just cleaning the lamps – washing, trimming the wicks, making them ready for the evening's lighting-up time.

Raffia, straw and basketwork shades should be vacuumed often and may be wiped with a damp cloth or rinsed out in detergent if you like, but make the solution weak and make sure you have rinsed it all off properly, because any left over detergent will be slightly sticky and attract more dust.

In kitchens, bathrooms and workrooms shades are more likely to get dusty and greasy than in other parts of the house. Wash shades and fittings more often in these areas, using a strong detergent if you think the fitting will stand up to it.

Don't forget the light bulbs themselves while you are cleaning the fittings. Take the bulb out of its socket first, unless you have chosen one of those dreadful fittings which it is impossible to get your hand into or out of, to put the bulb in. Dust regularly when you clean the room. Wipe once a month if you want to get full value from your fitting – making sure not to get the metal part wet, and dry the bulb thoroughly before putting it back.

Sun Lamps also need cleaning. But with these you should follow the manufacturers' instructions. Here are a few general hints: dust the reflector and keep it clean and polished, using suitable metal polish. Disconnect the lamp from the mains and make sure the bulb is cold. Wipe it carefully with a cloth damped with white spirit. Don't use soap or detergent and water, and don't handle it with bare fingers or you'll leave grease on the glass.

Old brass lamps may sometimes go a bit yellow. In that case wash them with gentle detergent and water solution, then use a spray polish and buff up well with a soft cloth. Many modern or reproduction lamps are already sprayed with a protective lacquer when they are new. You can lacquer your own lamps applying it with a camel-hair paint brush, but it may be difficult to get the lacquer – make sure the lamp is absolutely clean and shining before you start. The lacquer won't make the lamp shiny, it will simply preserve the appearance it already has. Slightly warm the lamp before applying the lacquer, this helps it to run on smoothly, like icing a cake with a knife dipped in hot water. The lamp must be scrupulously dry before you treat it. Some people prefer not to lacquer their brassware, in which case it will be necessary to use a proprietary brass polish to keep the lamp in good shine.

Maintenance
Check all fittings from time to time to see that the wires are secure, unfrayed and undamaged. Sometimes mice will get at them and chew them, sometimes age eats them away. If you are doubtful about whether the connections are good or not, check them out, don't wait until they cause a short circuit.

Keep a supply of bulbs in the linen cupboard or the larder. This is not such simplistic advice as you might suppose, since it is highly likely that several of the fittings in your home take different light bulbs and they may not be obtainable in every shop selling bulbs, so keeping a supply handy of every type you may need makes a lot of sense.

When fluorescent tubes start to flicker, it's time to change them. Such tubes fail gradually, not all at once, and poor starting, incomplete lighting or flickering all indicate that the tube is ready to be changed.

A selection of Christopher Wray shades

Christopher Wray makes traditional shades in many styles and sizes using the same equipment and techniques in the same factories as those in which they were originally made. This is a selection from his range.

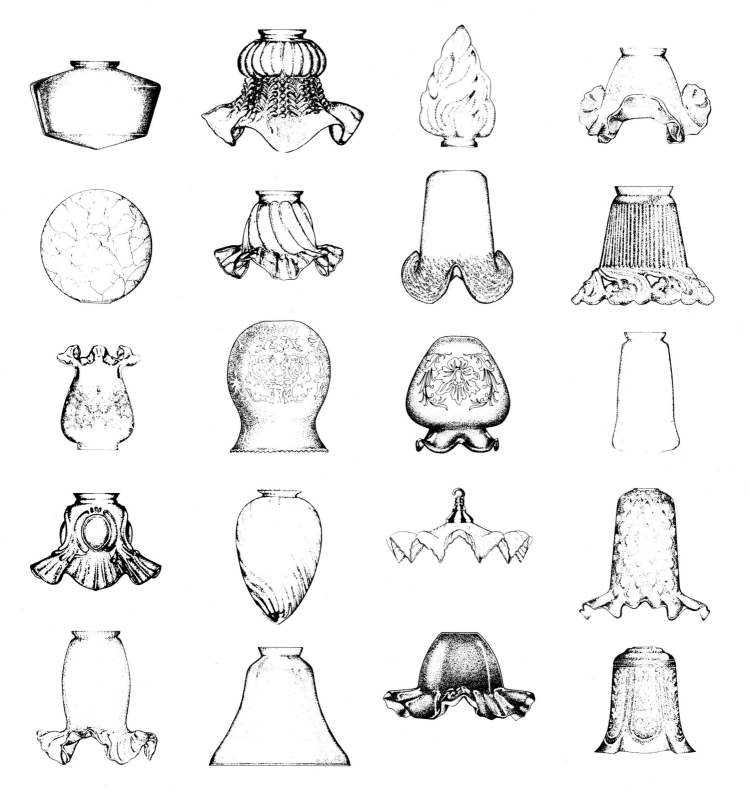

A selection of contemporary shades

Glossary of modern lighting terms

A-LAMP: standard light bulb (USA).

ACCENT LIGHTING: a narrow beam used to highlight a particular object or feature.

AMBIENT LIGHTING: see General Lighting

AMPERE (Amp): an internationally agreed unit of electric current.

ARC: a luminous discharge of electricity across a gap in an electric circuit or between two electrodes.

BACKGROUND LIGHTING: see General Lighting

BAFFLE: a shield attached to a light fitting to prevent glare.

BALLAST (or choke): a device to prevent fluorescent mercury-vapour and sodium bulbs etc from consuming more and more electric current, by allowing a high charge for starting the light and a low voltage for running it. Usually in a heavy black box.

BAYONET FITTING: standard type of bulb base with two ears for attaching a tungsten bulb to a lamp socket. Used in the UK.

BEAM: the line of light running from light source to object, particularly from a reflector bulb or fitting. The beam can be narrow, medium or wide.

BULB: the glass bubble that protects the light source. This is known as a 'lamp' in the trade.

CHOKE: see Ballast

CIRCUIT: the complete path of an electric current along the supply cables to light fittings and back to the beginning.

COMPACT FLUORESCENT: a small bulb such as the Philips PL and SL, and the Thorn 2D available in various shapes and operating on the same principle as the fluorescent tube. Can be used in a wide variety of fittings but cannot be dimmed.

COLD BEAM BULB: a type of PAR bulb with a dichroic reflector which greatly reduces the heat of the beam.

CROWN SILVERED BULB: a reflector bulb with an internally silvered reflective front which gives a sharply defined single beam.

DIFFUSED LIGHT: light filtered evenly through a translucent material.

DIRECT LIGHT: light provided directly from a fitting without using a reflector or reflecting off any other surface.

DOWNLIGHTER: a light fitting that casts light directly downwards.

EDISON SCREW: standard type of tungsten lamp base with a screw for attaching bulb base to lamp socket.

EYEBALL FITTING: semi-recessed downlighter which can be swivelled in its socket so that light can be directed at an angle.

FILAMENT: a thin wire (usually of tungsten) which emits light when heated to incandescence.

FITTING: the housing for a socket and bulb.

FLUORESCENT: white light in tubular glass case coated inside with phosphorescent powders. When switched on the phosphorus layer emits light.

FOOTCANDLE: unit for measuring light (USA).

GENERAL LIGHTING: a low level of light illuminating a whole area more or less uniformly.

GLARE: bright light shining into the eyes and impairing the vision.

GLS BULB: standard light bulb (UK).

INCANDESCENT LIGHT: yellow light produced by heating a material, usually tungsten, to a temperature at which it glows. Heat is given off as well as light.

INDIRECT LIGHT: light reaching its destination by being bounced off another surface.

ISL BULB: internally silvered reflector bulb, with silvering on the back of the bulb to reflect light forwards.

LAMP: in the trade this is the term for the tube or light bulb, but for most people it indicates the light fitting and base and that is how the term is used in this book.

LOUVERS: slats attached to a light fitting which can be angled to prevent glare.

LOW VOLTAGE BULB: a mini bulb running on 12 or 24 volts rather than mains voltage which is 240 volts in the UK and 110 in the USA. Such bulbs need a transformer to reduce the voltage.

LUMINAIRE: professional name for a fitting.

LUX: an internationally agreed unit used to measure the amount of light falling on a particular surface.

MERCURY VAPOUR BULB: a bulb which contains mercury vapour at low pressure and produces greenish blue light. Used for outdoor lighting.

METAL-HALIDE BULB: contains metal halides in a quartz discharge tube giving white light which reproduces colours accurately. Used for outdoor floodlighting.

MULTI MIRROR BULB: a miniature low voltage bulb with integral multi-faceted reflector. Used in many modern 'designer' lamps.

NEON LIGHT: a low pressure neon-filled tube which emits a reddish light. The name is given to similar tubes containing argon or krypton gases.

PAR BULB (parabolic aluminized reflector): a bulb which incorporates its own reflector to direct a powerful narrowish beam. Made of heat resistant glass.

PENDANT: a hanging ceiling light.

QUARTZ HALOGEN: see Tungsten Halogen

REFLECTOR BULB (UK): a variety of incandescent bulbs ranging from PAR 38, ISL or multi-mirror. They incorporate their own reflectors to direct light downwards. A scoop reflector is a broadly curved reflector in a wall, washing light which spreads the light to the top of the wall.

R-LAMP (USA): see Reflector Bulb

SHADE: a cover for a bulb to prevent glare, control light distribution and/or diffuse and colour the light.

SODIUM BULB: a bulb containing neon and sodium vapour at low pressure, producing orange coloured street lamps.

SPOT LIGHT: a single light source producing a directional beam.

TASK LIGHT: lighting to work or read by.

TRACK: an insulated fitting in various lengths into which light fittings can be clipped. Very sophisticated versions are available to accommodate up to four different circuits.

TRANSFORMER: a device to lower the domestic electricity supply from mains voltage to that suitable for low voltage lighting.

TUNGSTEN: see Incandescent

TUNGSTEN STRIP LIGHT: incandescent light in the form of a tube rather than a bulb, sometimes known as 'architectural tube'.

TUNGSTEN HALOGEN: a conventional incandescent filament containing a halogen gas. The gas combines with tungsten to create a far brighter light than equivalent wattage incandescent bulbs and prolongs the life of the bulb (also referred to as quartz halogen).

UPLIGHTER: a light fitting that casts its light upwards.

VA RATING: measure of the capacity of a transformer in terms of the wattage of the bulbs that may be connected to it.

VOLT: a unit expressing the potential of an electric circuit.

WALL WASHER: a lighting fixture that will cast light along and down a wall.

WATT: a unit of power describing the electrical output of a bulb.

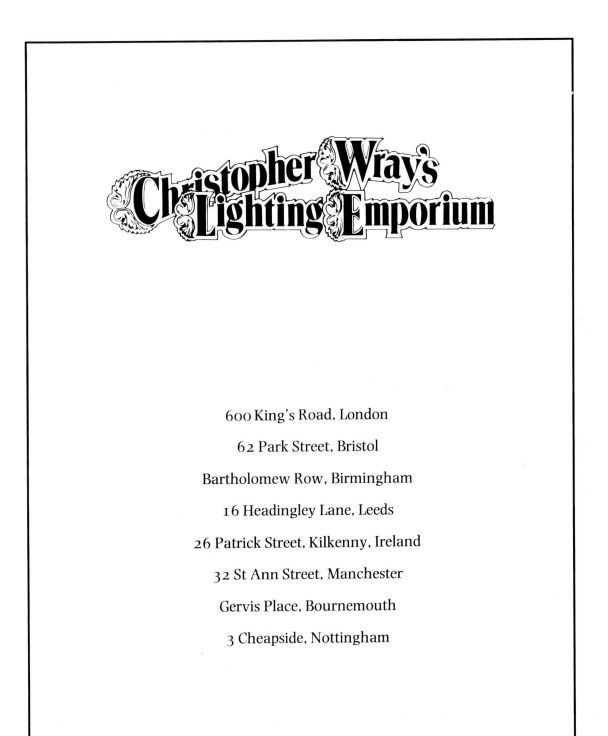

Christopher Wray's Lighting Emporium

600 King's Road, London

62 Park Street, Bristol

Bartholomew Row, Birmingham

16 Headingley Lane, Leeds

26 Patrick Street, Kilkenny, Ireland

32 St Ann Street, Manchester

Gervis Place, Bournemouth

3 Cheapside, Nottingham

Other useful addresses

BRITISH STANDARDS INSTITUTION (BSI), 2 Park Street, London W1H 2BS.
ELECTRICAL CONTRACTORS' ASSOCIATION (ECA), Esca House, 34 Palace Court, London W2 4HY.
BRITISH ELECTROTECHNICAL APPROVALS BOARD (BEAB), Mark House, The Green, 9–11 Queen's Road, Hersham, Walton-on-Thames, Surrey, T12 5NA.

LIGHTING SUPPLIERS

ARAM DESIGNS, Keane Street, London EC2.
Furniture and lighting showroom and design service.
ARGON NEON LTD, 27 Neal Street, London WC2.
Design and retail neon lights.
ARTEMIDE GB LTD, 17 Neal Street, London WC2.
Retail shop, selling upmarket Italian lighting.
CO-EXISTENCE, Whitcomb Street, London WC1.
Retail shops selling selection of Italian lighting.

JOHN CULLEN LIGHTING DESIGN, Woodfall Court, London SW3.
Design service and shop.
ERCO LIGHTING LTD, 38 Dover Street, London W1X 3RB.
Mainly contract lighting, but sell good holders for fluorescent.
FUTON COMPANY LTD, 654a Fulham Road, London SW6.
Retail shop with selection of Japanese-type table, floor and hanging lamps.
JOHN LEWIS, Oxford Street, London W1 and branches.
Lighting department has good selection of most types of freestanding and pendant lighting at reasonable prices.
THE LIGHTING WORKSHOP, 35–36 Floral Street, London WC2.
Lighting design service and showroom.
LIGHTSTYLE, 94 Tottenham Court Road, London W1.
HABITAT, Tottenham Court Road, London W1 and branches.
Reasonably priced and quite comprehensive lighting department.

HEALS, Tottenham Court Road, London W1.
Good looking selection of lighting.
LONDON LIGHTING CO. LTD, 135 Fulham Road, London SW3.
Retail shop with excellent selection of lighting of all kinds.
MR LIGHT, 307 Brompton Road, London SW3.
Retail shop with good selection of lighting.
NEW LIGHTING CENTRE, 62 Scotswood Road, Newcastle.
Retail shop and consultancy.
NOVA LIGHTING, 71 Candleriggs, Glasgow.
Retail shop and consultancy.
ROSET, 130 Shaftesbury Avenue, London W1.
Retail furniture shop, coloured fluorescents.
ROTAFLEX HOMELIGHTING, Malmesbury, Wilts SN16 0BN.
Specialise in spotlights and downlighters.
WEST MIDLAND LIGHTING CENTRE, 10–12 York Road, Erdington, Birmingham B23 6TE.
Retail shop.

Index

Acknowledgements

My grateful thanks to John Phillips for help in researching this book and to the following for inspiration, patience and in many cases the use of their homes for photography:

Lesley Aaronson, Simon and Alison Adams, Dorothy Block, Concord Lighting, Barry Daniels and Sue Donovan, Marianne Davies, Julia Fairhurst/Design Studio, Bronwen Glynn, Colin Huntley, Lars and Jo Huntley, Ligne Roset in Shaftesbury Avenue, David Rosenthal, Annabel and Mike Manwaring, Nick Meers, Peter and Ray Meers, Charlie Phillips, Martin Priestman, Sue Seddon, Angela Silverman, Mary Wiggin/Co-Existence in London and Bath, Christopher Wray, Lynda Young.

All of the photographs were taken by Nick Meers except: Paul Beattie 39; Ray Gaffney 16, 18, 22, 118; Concord Lighting 53; A Kolesnidow (room designed by Fiona Campbell for Pallu and Lake) 45; Christopher Wray Frontispiece, 59, 67, 73, 86.